Key to

Algebra®

5-7

Answers & Notes

By Julie King and Peter Rasmussen

KEY CURRICULUM PRESS
Innovators in Mathematics Education

Dear Teacher,

Too many students end their study of mathematics before ever taking an algebra course. Others attempt to study algebra, but are unprepared and cannot keep up. *Key to Algebra* was developed with the belief that *anyone* can learn basic algebra if the subject is presented in a friendly, non-threatening manner and someone is available to help when needed. Some teachers find that their students benefit by working through these books *before* enrolling in a regular algebra course–thus greatly enhancing their chances of success. Others use *Key to Algebra* as the basic text for an individualized algebra course, while still others use it as a supplement to their regular hardbound text.

Books 1-4 are restricted to operations on integers so that students who have not yet mastered fractions can begin their study of algebra. In Books 5-7 operations on fractions are taught as students study rational algebraic expressions. Books 8-10 cover real numbers, graphs, systems of equations, square roots and quadratic equations. Together these books comprise a complete introductory algebra course. Below are some suggestions for their effective use.

Allow students to work at their own pace. The *Key to Algebra* books are informal and self-directing. We suggest that you allow each student to proceed at his or her own pace. This will free you to work with individual students who need special help.

These workbooks are the written component of your program. They shouldn't replace group exploration of concepts, cooperative problem solving or class discussions. Use some of your class time for activities which build and reinforce algebraic concepts.

Use pencil only. Ask your students to use a pencil and not a pen as they work through the series. Encourage them to go back and correct their mistakes.

Keep a calculator in your classroom. Several pages in the series require or suggest the use of a calculator. You should keep at least one calculator in your classroom to loan to students who do not have their own.

Let students check their own work. Each problem in this answer book has been worked out just as if done by a student. You may want to let your students check their own booklets. If so, recommend that they stop and check their work every five pages or so. Students can use the answer book to mark problems wrong, but then should go back without the answer book to correct their mistakes. Don't consider the book complete until all mistakes have been found and fixed.

Use practice tests as a review. Each *Key to Algebra* book ends with a practice test that serves as a cumulative review for that booklet. Check the practice tests yourself. Go over the mistakes on the practice test individually with students if possible. If necessary, ask students to review before going on to the next booklet.

Discuss the covers. The covers of the *Key to Algebra* books trace the history of algebra. On the contents page of each book is an explanation of the cover illustration. Read these explanations aloud and discuss them with your students. You might ask students to read these again when they arrive at topics mentioned in the cover stories.

Read the notes on the following pages. They have been written to help you anticipate and resolve special problems that may arise. If you have any specific questions that are not answered in these notes or comments regarding *Key to Algebra* and its use, feel free to write us.

We hope that you and your students enjoy using *Key to Algebra*.

Julie King and Peter Rasmussen

Cover Illustration: Jay Flom
Graphic Design: Eleanor Henderson, Ann Rothenbuhler, Alen Wey, and Felicia Woytakv

Published by Key Curriculum Press, 1150 65th Street, Emeryville, CA 94608 http://www.keypress.com e-mail: editorial@keypress.com
Printed in the United States of America
11 12 13 BRP 25 24 23
ISBN 978-1-55953-014-9

Notes ◆ Book 5

Page 1: We do not expect that students will remember everything they have learned about fractions and decimals, but we have not devoted a section exclusively to reviewing this. Instead, the arithmetic of positive rational numbers is reviewed as we introduce operations with rational expressions. For many students, encountering rational number operations in this new context will be enough to refresh their memories. If you have students who seem to be struggling, you might give them short diagnostic tests such as those found in the *Key to Fractions* and *Key to Decimals* Reproducible Tests books. Then use *Key to Fractions* and *Key to Decimals* for whatever remedial work seems called for.

The word "fraction" will be used in this book to refer to anything written in the form $\frac{a}{b}$, whether a number or an algebraic expression. The terms "rational number" and "rational expression" will be used to distinguish between the two.

Pages 2-3: We have chosen $\frac{^-a}{b}$ as the standard form of a negative rational number (*a* and *b* here represent positive integers) to avoid confusion and to make operations with fractions simpler. The problem $\frac{4}{5} + \frac{1}{^-5}$, for example, can immediately be rewritten as $\frac{4}{5} + \frac{^-1}{5}$.

Using this form, the opposite of $\frac{a}{b}$ will be written as $\frac{^-a}{b}$ and so, of course, the opposite of $\frac{^-a}{b}$ is $\frac{a}{b}$. This makes subtraction of rational numbers easy. To subtract $\frac{^-2}{3}$, we add its opposite, $\frac{2}{3}$. The subtraction problem $\frac{5}{3} - \frac{^-2}{3}$ is rewritten as $\frac{5}{3} + \frac{2}{3}$.

This gives us a simple way of dealing with what are traditionally called "the three signs of a fraction." Later, in Book 7, we will use a similar approach in dealing with opposites of rational expressions.

Page 4: Remind students that we can cancel, as in the example, because we know that when a rational number is multiplied and then divided (or divided and then multiplied) by the same number, the result is the number we started with. In other words, division "undoes" multiplication. In Book 6, after we have introduced multiplication of rational numbers, we will discuss canceling as eliminating a factor of 1.

Page 6: If students are having trouble with problems on this page, you can ask them to list a few integers which fit the description before graphing the set.

Page 7: A mistake students often make in graphing rational numbers is to count to the *right* instead of to the *left* of the integer when graphing a negative mixed number. Watch for this.

Page 8: You may want to supplement this page with a worksheet which requires students to subdivide the units.

Page 9: The number line contains rational and irrational numbers (such as π and $\sqrt{3}$). Thus the shading includes, but is not limited to, all the rational numbers.

Page 11: A student who makes more than one or two errors on the true-false problems may not be interpreting the symbols correctly. Ask the student to read aloud each true or false statement in the sections above, and to explain why it is true or false.

Pages 12-13: Each of the new and more involved symbols has synonyms. Another way to say "is not greater than or equal to" would be "is less than," for example. Using a simpler expression will make the last three problems on page 13 easier.

Page 14: Caution: It is tempting to conclude that since one solution of $|x + 2| = 4$ is 2, the other solution will be $^-2$. Not so, as students will see if they try substituting $^-2$ in the equation. The other solution is $^-6$. $|x + 2| = 4$ is actually equivalent to the *pair of equations* $x + 2 = 4$ or $x + 2 = {^-4}$. Encourage students to think of all the possible values of the expression within the absolute value sign, then to decide what to substitute for x in order to obtain these values. The mechanical process should not be emphasized here. It is the thought process which is important for students to understand.

The problems in the last section require some concentrated thought. Again, reading the inequalities aloud and trying specific numbers may be helpful. Do not let students rush through this section.

Pages 15-19: There are many details to pay attention to on these pages. Haste will not give students any advantage here. The inequalities on pages 18 and 19 serve as a review of the techniques for solving linear equations. If students have forgotten some of these, they should go back to study and review them. The last six problems look like quadratic inequalities, but are actually linear. They are easily solved if the second degree terms are subtracted from both sides.

Pages 20-23: It is easy to confuse multiplying or dividing by a negative number with getting a negative result. If students do this, focus their attention by asking what number they multiplied or divided *by.*

Pages 24-26: On these pages we make explicit the process of finding two equations equivalent to one which contains the absolute value of an expression. You can reinforce this by advising the students to STOP — LOOK BOTH WAYS whenever they reach the stage at which the absolute value of an expression is equal to a number. This is the point where a pair of equivalent equations must be written.

Pages 27-28: The verbal translations on these pages may seem a little awkward. It is more natural to say "Helen is a sister of Guy" than to say "A sister of Guy is Helen." We have used the latter because it follows the same order as the symbols in the functional equation.

Page 31: Leap years are all non-centennial years which are divisible by 4, and all centennial years divisible by 400.

1600 was a leap year.
1700, 1800 and 1900 were not leap years.
2000 will be a leap year.

Students may enjoy suggesting and naming some other specific functions they use frequently, such as the "Change from $1.00 Function." Examples of everyday functions would make a good display.

Page 34: Although the formulas for these functions seem simple once they are known, some of them may be difficult to guess. Don't allow students to give up easily or look at the answers too quickly. The rewards of having persisted and found a solution will be satisfaction and increased self-confidence. Most students will be able to figure out the formulas eventually, even it they have to "sleep on it."

Pages 35: The Written Work page provides a review of some of the concepts covered in this book. It is also designed to develop the skills needed in preparing the kind of written assignments required of students who are using a regular math textbook. Some of the problems demand that students reorganize their knowledge or apply it in a new way, and will not be exactly like problems they have done previously.

Pages 36-37: The Practice Test can be used as an informal check and diagnostic tool for the student and teacher to determine whether the student has mastered the material in Book 1. Students who "pass" the Practice Test can be given a more formal teacher-made final test or the Final Test for Book 5 from the Reproducible Test booklet, and go on to Book 6 if they pass it. When students do poorly on the Practice Test or Final Test, you should work with them individually in areas where they need help. For key topics, it is helpful to maintain a file of self-made supplementary worksheets, prepared in advance, to use in such instances. With this approach, your students should have all the knowledge and confidence they need to get off to a good start in Book 6.

Notes ◆ Book 6

Page 1: A sound understanding of all the processes reviewed on this page will make the work in Book 6 much easier, as it will free students to concentrate on the ideas which are new. Page 1 should help you to spot areas which individual students need to review before proceeding. You can refer them to the following pages in previous books for particular topics:

	Book	Pages
Writing rational numbers as fractions	5	1
Graphing rational numbers on a number line	5	7-9
Using exponential notation	2	6-9
Multiplying powers	2	12-14
Multiplying a monomial by a binomial	4	7-8
Multiplying two binomials	4	15-17
Factoring out a common factor	4	11-14
Factoring trinomials	4	19-28
Solving equations	3	32-34

It will be helpful to have a prepared worksheet on each topic to hand students after they have reviewed a process.

Page 2: The same polynomial may be used as both the numerator and denominator of rational expressions the students create.

$$\frac{x^2}{x^2} \qquad \frac{10}{10} \qquad \frac{x-5}{x-5} \qquad \frac{x^2+3x+2}{x^2+3x+2}$$

Some students may recognize that each of these is equivalent to 1.

Page 3: Watch out for mistakes in squaring $^-2$.

Pages 4-5: We want students to be aware that substitutions which result in 0 denominators are not permissible, without dwelling on this in great detail at this point. It is not necessary to ask students to determine *which* values of x would give 0 denominators.

Page 6: This page serves to review multiplication of rational numbers as well as to introduce multiplication of rational expressions. If you want to extend the review, you can try this instead of providing additional pages for drill: Ask students to rewrite the numerical problems as they might look on a page from an arithmetic book. For example,

$\frac{5}{3} \cdot \frac{7}{2}$ might appear as $1\frac{2}{3} \cdot 1\frac{1}{2}$.

Pages 9-13: The process of finding an equivalent fraction in higher terms will not be needed until Book 7, but we deal with it here in detail so that students will be able to see that simplifying a fraction is the reverse process. Students are less likely to cancel inappropriately when simplifying if they understand that canceling really amounts to eliminating a factor of 1.

On page 13 we have left the denominators in factored form because this will be the most convenient form to use for addition and subtraction. It is not incorrect, of course, for students to carry out the multiplication in the numerator and denominator.

Page 15: Students should write out all factors before canceling to assure that they do not fall into the error of canceling only the base and leaving the exponents. If they find this tedious, assure them that they will not have to do it forever.

Page 16: Most of these problems are identical to the problems the students did on page 15. Now they have a chance to try doing them mentally. Ask those having difficulty to look back at how they solved the same problems on page 15.

Page 18: In some of these problems the common factors of the numerator and denominator are binomials. Send students back to page 12 to look at the reverse process if they are having trouble.

Pages 19-21: Check often to make sure students are canceling *factors*, not *terms*.

Pages 22-23: We have avoided introducing complications here, such as missing terms in the dividend and division by polynomials of degree higher than 1. However, division of one polynomial by another is a lengthy procedure which may appear daunting to some students. Your support and encouragement may be needed. Most errors will probably be in subtracting polynomials.

Page 25: Students will be most successful in simplifying problems which involve binomials or trinomials if they first factor out the greatest common factor of the terms.

Page 27: Emphasize factoring *before* simplifying as well as simplifying *before* multiplying.

Page 29: To help focus attention on the division process, you can ask students to do the following: Go down the left column and rewrite the first fraction in each problem; next write a multiplication sign in place of each division sign; third, write the reciprocal of each second fraction; finally, do the multiplication to find the answer.

Page 30: We have deliberately put the rule at the bottom of the page to give students a chance to figure it out for themselves.

Page 32: Make sure students simplify only *after* writing an equivalent multiplication problem.

Pages 35-37: See comments for Book 5.

Notes ◆ Book 7

Page 1: Students who have just finished Book 6 should have no trouble with most of the problems on this page unless they are uncertain about addition, opposites, and subtraction of polynomials. If so, they should review these topics in Book 4 before proceeding.

Page 2: As in Book 6, operations with rational numbers are reviewed in the process of introducing the same operations with rational expressions. Students may not be accustomed to seeing addition problems written horizontally. To get them off to a good start, write the first problem as

$$\begin{array}{r} \frac{5}{8} \\ +\ \frac{2}{8} \\ \hline \end{array}$$

Ask the students to write the next two problems this way also, and to do all three of these problems on a separate piece of paper before beginning the work in the book.

Pages 3-5: Keep a close eye on students' work to make sure they are remembering to factor *before* simplifying.

Page 6: Students may be able to see immediately that the answers to the last four problems will all be 0. If not let them write out all the steps so that they have a chance to discover for themselves that opposites add up to 0.

Page 7: If students have difficulty with these, ask them to apply steps 1 and 2 to several problems in succession before finding any answers.

Page 8: Some students will find this page easier if they do all the subtracting first, and then return to the answers to see whether any can be simplified.

Pages 9-10: Knowing when it is more convenient to leave expressions in factored form and when they need to be multiplied out is a sense which will develop as students gain facility with addition and subtraction. Here we give instructions, but later we will leave this to the student to decide.

Page 11: Ask students who have difficulty with these problems to write "thought bubbles" for the problems which have been worked out. From the "bubbles" you will be able to tell whether they have studied and understood the examples.

Page 12: These problems require a careful and methodical approach. Encourage students to think through each step. Ask leading questions if necessary: "Can you multiply one denominator by something to get the other? How can you find out what to multiply by?" Completing the page may take time, but most students will take satisfaction in having finished it successfully.

Page 13: Students may not immediately recognize that because of the Commutative Principle, $2x + 5$ is equivalent to $5 + 2x$. Having realized this, they may also conclude that $x - 3$ is equivalent to $3 - x$. Remind them that $x - 3$ is equivalent to $x + {}^-3$ and therefore to ${}^-3 + x$, not to $3 - x$. They should then be able to recognize that $x - 3$ is the *opposite* of $3 - x$.

If students do not realize it themselves, also point out that multiplying a polynomial by ${}^-1$ will change the sign of each term, so that the result is the opposite of the polynomial, as in the second example at the top of the page.

Page 14: Students might want to know why we can't write $1\frac{3}{x}$ to stand for $1 + \frac{3}{x}$. This would be a good topic for discussion and debate.

Page 16: As on page 2, it may help some students to write out the first four problems in vertical form before doing them horizontally, in order to make connections with what they already know about adding and subtracting fractions.

Pages 18-19: A method for finding a least common denominator is given on page 19. On the previous page we want students to experiment, and to form their own rules if possible.

Pages 20-22: See comments for page 12.

Pages 23-24: Equations obtained by multiplying both sides of a given equation by an expression containing a variable may not be equivalent to the original equation. For this reason, an equation containing rational expressions may have extraneous solutions — solutions of the final equation which are not solutions of the original equation. None of the equations in this book has extraneous solutions. We have chosen not to discuss extraneous solutions in Book 7 in order to allow students to concentrate on learning the process. Instead we have simply modeled checking the solutions.

Page 26: Students sometimes refer to multiplying by the reciprocal in a division problem as "cross multiplying." We prefer to reserve the term for the process used in solving a proportion.

Page 27: The last six problems result in quadratic equations. Students should review the method of solving quadratic equations by factoring (Book 4, pages 30-32) if they have forgotten it.

Pages 28-29: It may help students to write out the quantities being compared in words before writing the proportion. For example, $\frac{\text{gallons}}{\text{miles}} = \frac{\text{gallons}}{\text{miles}}$

Pages 30-31: Textbooks sometimes advise students to set up a proportion involving a percentage by using the formula

$$\frac{\text{part}}{\text{whole}} = \frac{\text{percent}}{100}$$

This is helpful in some cases, but can be confusing when the "part" is greater than the "whole" (i.e. the percent is greater than 100), as in the first problem about Plainview High on page 31.

Pages 32-33: The problems on these two pages are traditionally called "work problems," but the thought process can be applied to other situations as well (for example the dog food problem on page 33). Watch out for the duplicator problem. The times are expressed in different units.

Pages 35-37: See comments for Book 5.

Key to Algebra – ANSWERS

Book 5, Page 1

Rational Numbers

In Books 1 to 4 we worked with integers (positive and negative whole numbers and 0). We had no trouble adding, subtracting and multiplying integers, but when we got to division we ran into difficulties. Division problems like $9 \div 0$ have no answer, because you can never divide by 0. Other problems, like $10 \div 3$, do not have answers which are integers.

To solve problems like $10 \div 3$ we need a new class of numbers called **rational numbers**. Rational numbers are numbers which can be written as fractions. The **numerator** (top number) and **denominator** (bottom number) of a fraction must be integers and the denominator may not be 0.

$\frac{3}{4}$ $\frac{1}{7}$ $\frac{-5}{2}$ $\frac{0}{8}$ $\frac{-6}{11}$ $\frac{-9}{1}$ $\frac{-7}{3}$ ← numerators ← denominators

Every integer is a rational number because it can be written as a fraction with a denominator of 1. Rewrite each integer as a fraction.

$8 = \frac{8}{1}$ $\quad -3 = \frac{-3}{1}$ $\quad 0 = \frac{0}{1}$ $\quad -15 = \frac{-15}{1}$

$4 = \frac{4}{1}$ $\quad -21 = \frac{-21}{1}$ $\quad 25 = \frac{25}{1}$ $\quad 6 = \frac{6}{1}$

Every mixed number is a rational number because it can be written as a fraction. Rewrite each mixed number as a fraction.

(2·8=16. 2 is 16 eighths and there are 3 more eighths.)

$2\frac{3}{8} = \frac{16+3}{8} = \frac{19}{8}$ $\quad 1\frac{3}{4} = \frac{4+3}{4} = \frac{7}{4}$ $\quad 10\frac{2}{3} = \frac{30+2}{3} = \frac{32}{3}$

$5\frac{1}{2} = \frac{10+1}{2} = \frac{11}{2}$ $\quad 7\frac{1}{7} = \frac{49+1}{7} = \frac{50}{7}$

$3\frac{2}{5} = \frac{15+2}{5} = \frac{17}{5}$ $\quad 4\frac{5}{6} = \frac{24+5}{6} = \frac{29}{6}$ $\quad 3\frac{3}{7} = \frac{21+3}{7} = \frac{24}{7}$

Every decimal is also a rational number (unless it goes on forever without repeating). A **terminating decimal** (one that comes to an end) is a rational number because it equals a fraction with a denominator of 10 or 100 or 1000, etc. Rewrite each terminating decimal as a fraction.

$0.6 = \frac{6}{10}$ $\quad 0.9 = \frac{9}{10}$ $\quad 1.3 = 1\frac{3}{10} = \frac{13}{10}$

$0.06 = \frac{6}{100}$ $\quad 0.09 = \frac{9}{100}$ $\quad 2.7 = 2\frac{7}{10} = \frac{27}{10}$

$0.006 = \frac{6}{1000}$ $\quad 0.19 = \frac{19}{100}$ $\quad 5.01 = 5\frac{1}{100} = \frac{501}{100}$

$0.056 = \frac{56}{1000}$ $\quad 0.119 = \frac{119}{100}$ $\quad 3.27 = 3\frac{27}{100} = \frac{327}{100}$

Book 5, Page 2

Dividing Integers

Now we can divide any integer by any other integer except 0. All we have to do is write a fraction with the dividend (the number we are dividing *into*) as the numerator (top) and the divisor (the number we are dividing *by*) as the denominator (bottom).

$$10 \div 3 = \frac{10}{3}$$

Do each division problem. If the divisor goes evenly into the dividend, write your answer as an integer. Otherwise, write it as a fraction.

$12 \div -2 = -6$ $\quad 15 \div 3 = 5$ $\quad 15 \div 4 = \frac{15}{4}$ $\quad -8 \div -3 = \frac{8}{3}$

$40 \div 7 = \frac{40}{7}$ $\quad -7 \div 2 = \frac{-7}{2}$ $\quad 1 \div -6 = \frac{1}{-6}$ $\quad 54 \div 7 = \frac{54}{7}$

$-3 \div 5 = \frac{-3}{5}$ $\quad -45 \div -9 = \frac{45}{9}$ $\quad -7 \div 6 = \frac{-7}{6}$ $\quad -2 \div 19 = \frac{-2}{19}$

A fraction can be positive or negative. To find the sign, just follow the rules for division. When division is written using a fraction bar, the rules look like this:

$\frac{\text{POSITIVE}}{\text{POSITIVE}}$ = POSITIVE $\qquad \frac{\text{NEGATIVE}}{\text{POSITIVE}}$ = NEGATIVE

$\frac{\text{POSITIVE}}{\text{NEGATIVE}}$ = NEGATIVE $\qquad \frac{\text{NEGATIVE}}{\text{NEGATIVE}}$ = POSITIVE

If a fraction is positive, we will write it with no signs. If a fraction is negative, we will write it with the negative sign on top.

$\frac{-3}{-4}$ is positive, so we will write $\frac{3}{4}$. $\qquad \frac{3}{-4}$ is negative, so we will write $\frac{-3}{4}$.

Do each division problem. Write your answer as a positive or negative fraction.

$13 \div -3 = \frac{-13}{3}$ $\quad -4 \div -7 = \frac{4}{7}$ $\quad 15 \div -2 = \frac{-15}{2}$ $\quad 18 \div -5 = \frac{-18}{5}$

$-3 \div 10 = \frac{-3}{10}$ $\quad 1 \div -5 = \frac{-1}{5}$ $\quad -1 \div 9 = \frac{-1}{9}$ $\quad 40 \div 29 = \frac{40}{29}$

$-9 \div -10 = \frac{9}{10}$ $\quad 9 \div -5 = \frac{-9}{5}$ $\quad 12 \div -7 = \frac{-12}{7}$ $\quad -3 \div -100 = \frac{3}{100}$

$12 \div 11 = \frac{12}{11}$ $\quad 4 \div 5 = \frac{4}{5}$ $\quad -9 \div -14 = \frac{9}{14}$ $\quad -20 \div 7 = \frac{-20}{7}$

Book 5, Page 3

Divide. Write your answer as an integer or as a positive or negative mixed number.

(4 goes into 9 2 times with a remainder of 1.) (7 goes into 12 1 time with a remainder of 5. The answer is negative.)

$9 \div 4 = 2\frac{1}{4}$ $\quad -12 \div 7 = -1\frac{5}{7}$ $\quad -18 \div -7 = 2\frac{4}{7}$

$50 \div -5 = -10$ $\quad 30 \div 9 = 3\frac{3}{9}$ or $3\frac{1}{3}$ $\quad -7 \div -6 = 1\frac{1}{6}$

$28 \div -4 = -7$ $\quad -18 \div 5 = -3\frac{3}{5}$ $\quad -60 \div 4 = -15$

$-80 \div -10 = 8$ $\quad 46 \div -2 = -23$ $\quad 64 \div -3 = -21\frac{1}{3}$

$25 \div -7 = -3\frac{4}{7}$ $\quad -63 \div -9 = 7$ $\quad -100 \div 3 = -33\frac{1}{3}$

$0 \div -6 = 0$ $\quad 20 \div 3 = 6\frac{2}{3}$ $\quad -14 \div -5 = 2\frac{4}{5}$

$-11 \div 4 = -2\frac{3}{4}$ $\quad -25 \div -4 = 6\frac{1}{4}$ $\quad -37 \div 10 = -3\frac{7}{10}$

Divide. This time write your answer as a positive or negative decimal.

$-3 \div 10 = \frac{-3}{10} = -0.3$	$107 \div 100 = \frac{107}{100} = 1\frac{7}{100} = 1.07$
$-8 \div -10 = \frac{8}{10} = 0.8$	$-41 \div 100 = \frac{-41}{100} = -0.41$
$39 \div 10 = \frac{39}{10} = 3.9$	$-253 \div -100 = \frac{253}{100} = 2.53$
$15 \div 4 = 3.75$ ($4\overline{)15.00}$ = 3.75)	$-26 \div 5 = \frac{-26}{5} = -5.2$
$-17 \div 2 = \frac{-17}{2} = -8.5$	$30 \div -4 = \frac{-30}{4} = -7.5$
$-100 \div -8 = \frac{100}{8} = 12.5$	$-12 \div 5 = \frac{-12}{5} = -2.4$

Book 5, Page 4

Equations with Rational Solutions

Now we can use the Division Principle to solve equations even when the answer is not an integer. Solve each equation. Write your answer as a fraction or as a mixed number.

$7x = -25$; $\frac{7x}{7} = \frac{-25}{7}$; $x = -3\frac{4}{7}$	$9x = 40$; $\frac{9x}{9} = \frac{40}{9}$; $x = 4\frac{4}{9}$	$-4x = 17$; $\frac{-4x}{-4} = \frac{17}{-4}$; $x = -4\frac{1}{4}$
$2x - 5 = 14$; $\frac{2x}{2} = \frac{19}{2}$; $x = 9\frac{1}{2}$	$3x + 7 = -4$; $\frac{3x}{3} = \frac{-11}{3}$; $x = -3\frac{2}{3}$	$-5x + 1 = 15$; $\frac{-5x}{-5} = \frac{14}{-5}$; $x = -2\frac{4}{5}$
$x - 3 = 8x + 5$; $-3 = 7x + 5$; $\frac{-8}{7} = \frac{7x}{7}$; $x = \frac{-8}{7}$ or $-1\frac{1}{7}$	$-2(x - 5) = 7$; $-2x + 10 = 7$; $\frac{-2x}{-2} = \frac{-3}{-2}$; $x = \frac{3}{2}$ or $1\frac{1}{2}$	$x - 5x + 7 = -8$; $-4x + 7 = -8$; $\frac{-4x}{-4} = \frac{-15}{-4}$; $x = 3\frac{3}{4}$

Solve these equations, too. This time if your answer is not an integer, write it as a decimal.

$4x = -10$; $\frac{4x}{4} = \frac{-10}{4}$ ($4\overline{)10.0}$ = 2.5); $x = -2.5$	$-5x = 18$; $\frac{-5x}{-5} = \frac{18}{-5}$; $x = -3.6$	$5 \cdot \frac{-2x}{5} = 7 \cdot 5$; $\frac{-2x}{-2} = \frac{35}{-2}$; $x = -17.5$
$7x - 7 = 3x + 20$; $4x - 7 = 20$; $\frac{4x}{4} = \frac{27}{4}$; $x = 6.75$	$x - 9 = 6x + 7$; $-9 = 5x + 7$; $\frac{-16}{5} = \frac{5x}{5}$; $x = -3.2$	$5 \cdot \frac{10x - 2}{5} = 3 \cdot 5$; $10x - 2 = 15$; $\frac{10x}{10} = \frac{17}{10}$; $x = 1.7$
$3(x - 5) = x - 20$; $3x - 15 = x - 20$; $2x - 15 = -20$; $\frac{2x}{2} = \frac{-5}{2}$; $x = -2.5$	$4(x + 6) = 23$; $4x + 24 = 23$; $\frac{4x}{4} = \frac{-1}{4}$; $x = -0.25$	$2(x - 3) + x = 9$; $2x - 6 + x = 9$; $3x - 6 = 9$; $\frac{3x}{3} = \frac{15}{3}$; $x = 5$

Key to Algebra – ANSWERS

Book 5, Page 5

Number Lines

In Book 1 we used **number lines** to help us think about adding and multiplying integers. The football field was a kind of number line. Rulers and the scales on thermometers are also number lines.

To make a number line we draw a line and divide it into sections of equal length called **units**. Then we number the points which separate the units. Here are three number lines:

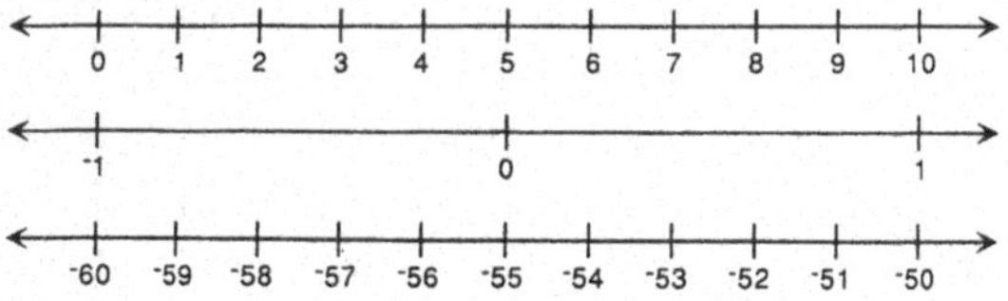

The arrows on the ends of each number line show that the number line keeps going. We can start with any number as long as we number the points in order (usually from left to right). Sometimes we do not show every unit. This number line only shows every fifth unit:

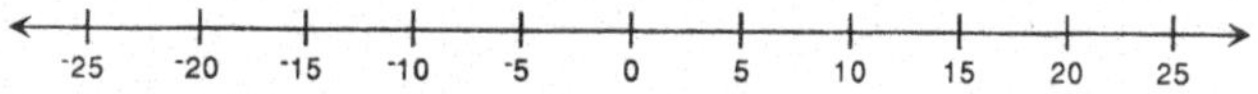

Here are some number lines for you to finish numbering:

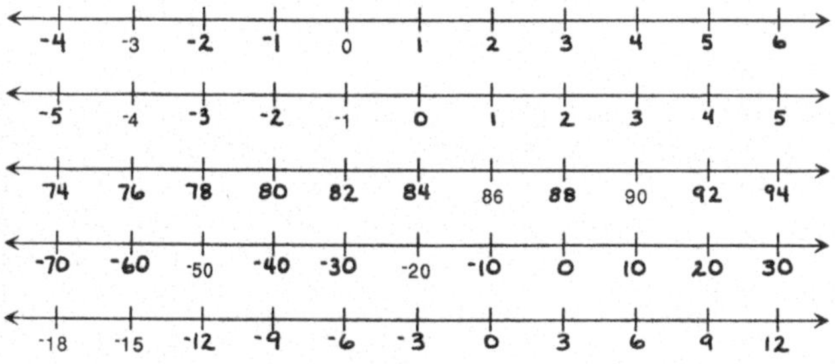

Make a number line showing all the integers from ⁻5 to 5.

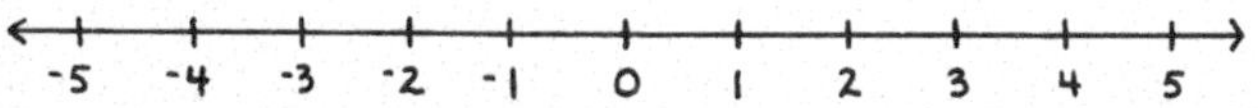

Book 5, Page 6

Graphing Integers

We can use a number line to picture a set of numbers. On the number line we make a dot to show each number in the set. This is called a **graph** of the set. A graph can help you see a pattern or answer a question. If a pattern continues forever to the left or right, we fill in the arrow that points in that direction.

Graph each set of integers below.

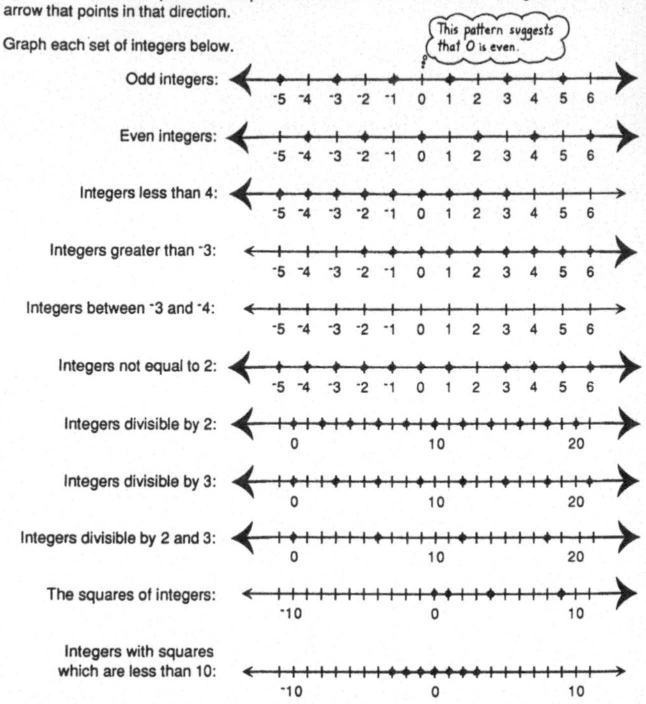

Did you notice any interesting patterns in the graphs you made?

Book 5, Page 7

Graphing Rational Numbers

Integers are not the only points on a number line. On the number line below we have also labeled the points halfway between each integer and the next.

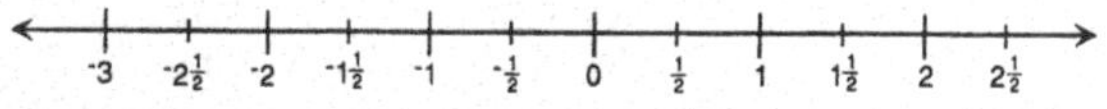

In fact, there is a point on the number line for *each* rational number. To find this point, first write the rational number as a fraction. The denominator of the fraction tells how many parts to divide each unit of the number line into. The numerator tells how many parts to count off to the right of 0 (if the number is positive) or to the left of 0 (if the number is negative) to find the point. Here's how to find $\frac{1}{6}$, $\frac{-3}{5}$, and $\frac{19}{7}$.

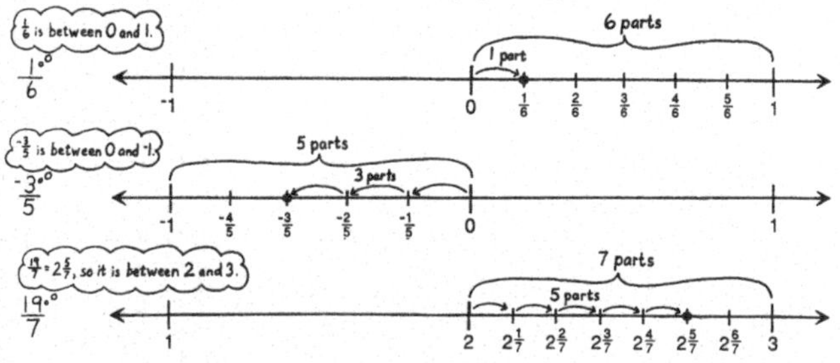

On each number line, first finish labeling the points. Then graph the rational number at the left.

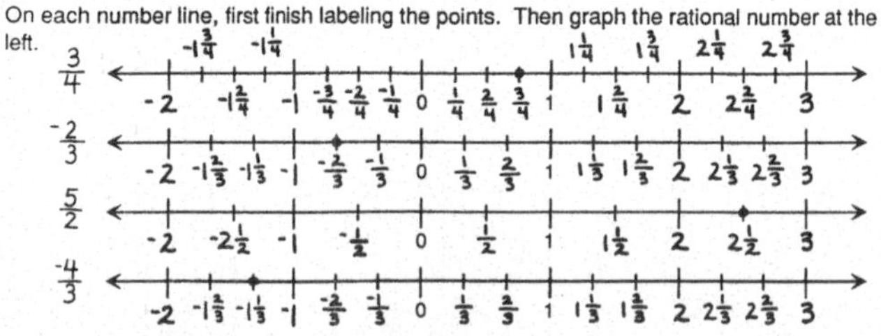

Label each rational number shown on the number line below.

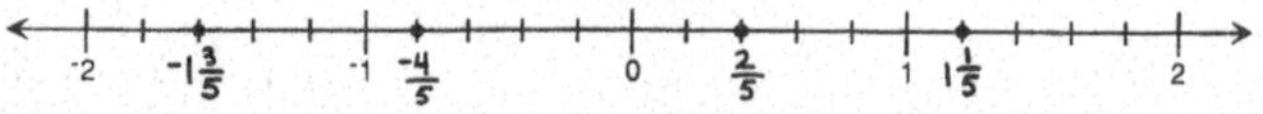

Book 5, Page 8

Each decimal is a rational number (unless it goes on forever without repeating), so it also has a place on the number line. To find the point for a decimal, think of it as a fraction or mixed number.

0.4 is the same as $\frac{4}{10}$. This number is between 0 and 1 so we divide that unit into ten parts and count four to the right of 0.

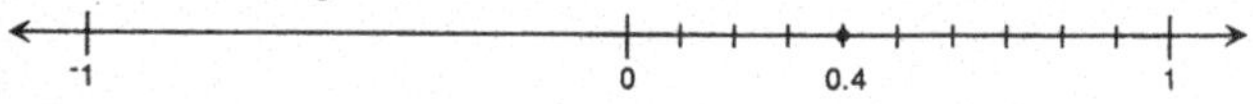

⁻3.7 is equal to $^{-}3\frac{7}{10}$. This number is between ⁻3 and ⁻4 so we divide that unit into ten parts and count seven units to the left of ⁻3.

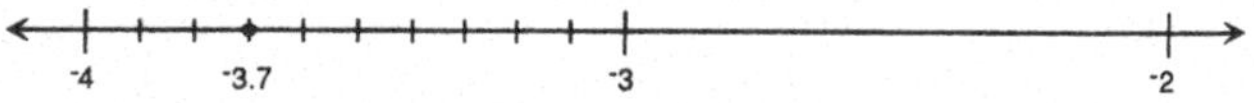

For hundredths we could divide the unit into a hundred parts, but to save time it makes sense to divide it into tenths first and then to divide only one of the tenths into ten parts. To graph 0.32 this way we first notice that it is between 0.3 and 0.4. Then we divide the section between 0.3 and 0.4 into ten parts. Each of these parts is a hundredth of the unit.

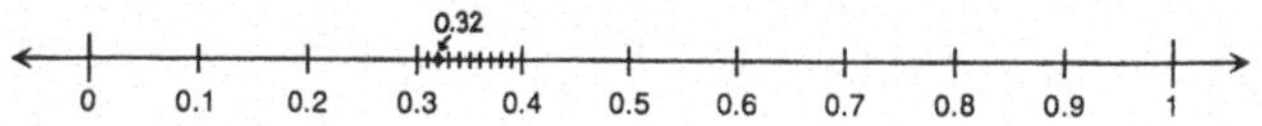

Graph each decimal below.

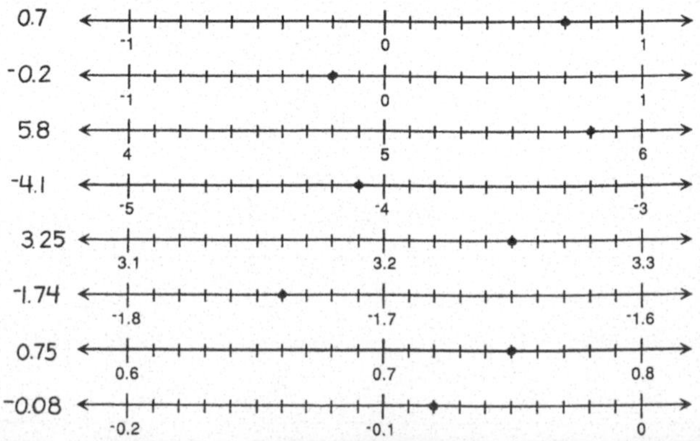

Key to Algebra – ANSWERS

Book 5, Page 9

Find a decimal name for each point graphed on the number lines below.

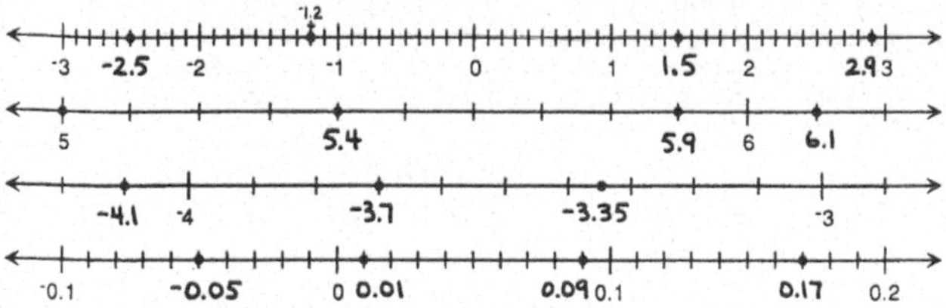

Imagine making a graph of all the rational numbers between 2 and 3.

First we would graph the halves,

then the thirds,

then the fourths,

then the fifths,

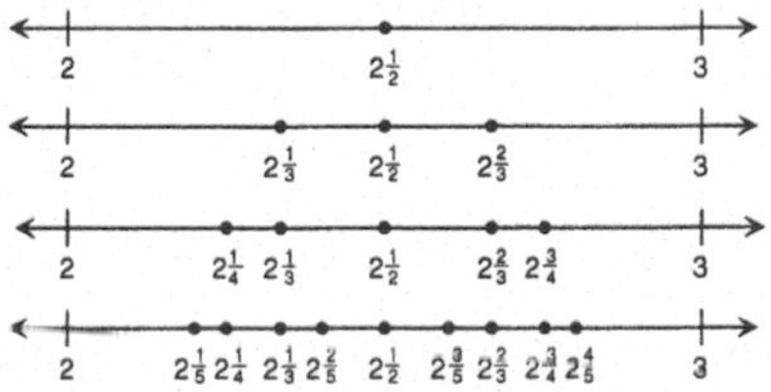

and so on . . .

We would never be finished! Soon the line would be so crowded with dots that you couldn't tell one from another. So when we want to show *all* the rational numbers between 2 and 3 we just shade the whole section of the line between those numbers.

Whenever we say "between" we will mean "not including the endpoints."
We have used hollow dots at 2 and 3 to show that those numbers are not included.

You graph all the rational numbers which are:

between ⁻1 and 4

between ⁻3 and 0

between 2 and 3.5

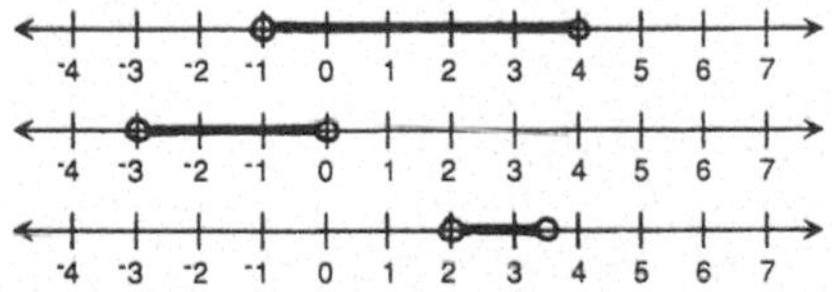

Book 5, Page 10

Can you tell what sets have been graphed?

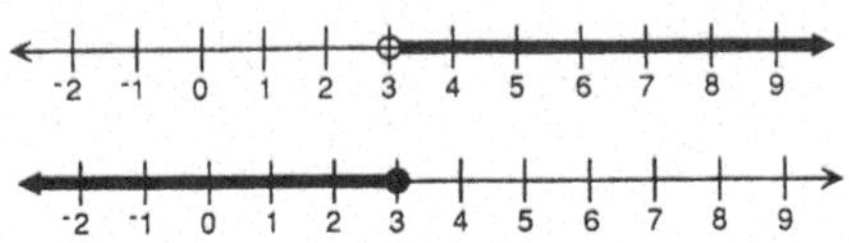

The first graph shows all rational numbers which are greater than. 3. The hollow dot shows that 3 is not included.

The second graph show all rational numbers which are less than or equal to 3. This time 3 *is* included, so we have used a solid dot. On both graphs the arrows have been filled in to show that the graphs continue.

Graph all the rational numbers which are:

less than 6

greater than 1

greater than or equal to 4

less than or equal to 0

greater than or equal to ⁻1.4

less than or equal to 0.5

between 8 and 8.5

not equal to 1

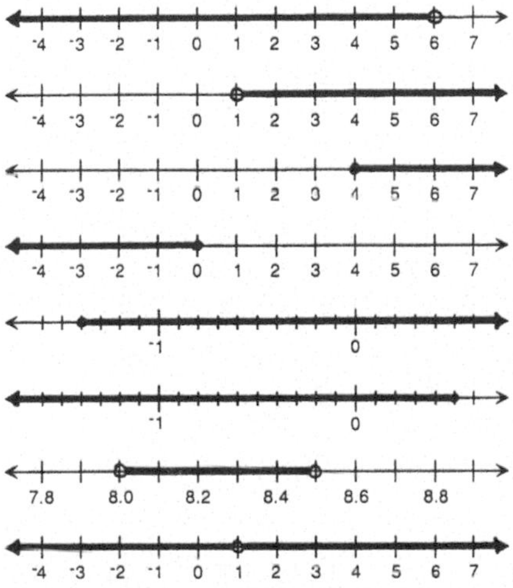

Book 5, Page 11

Inequalities

In Book 3 we worked with equations. Remember that an equation is a sentence about numbers being *equal*, like $x + {}^-4 = 5$.

Another kind of sentence is an **inequality** — a sentence about numbers being *unequal*. Here are two examples of inequalities:

$x + {}^-4 < 5$ means "$x + {}^-4$ *is less than* 5."

$x + {}^-4 > 5$ means "$x + {}^-4$ *is greater than* 5."

Sometimes we combine two symbols to make a new symbol. The symbol ≤ means "is less than or equal to." And the symbol ≥ means "is greater than or equal to." Sentences using these combined symbols are also inequalities.

$x + {}^-4 \le 5$ means "$x + {}^-4$ *is less than or equal to* 5."

$x + {}^-4 \ge 5$ means "$x + {}^-4$ *is greater than or equal to* 5."

To get used to using <, >, ≤ and ≥, read the following statements carefully.

Each of these is true:

10 > 6	5 ≥ 5	⁻8 < 8	12 = 12	⁻4 ≥ ⁻7
9 ≥ ⁻2	5 ≤ 5	8 > ⁻8	12 ≥ 12	⁻7 ≤ ⁻4

Each of these is false:

3 > 4	⁻6 < ⁻6	⁻9 ≥ 9	1 ≤ 0	1 ≥ 2
3 ≥ 4	⁻6 > ⁻6	9 ≤ ⁻9	0 < 0	2 > 2

Mark each statement below true (T) or false (F).

18 > 2 T	⁻14 ≤ 1 T	7 < 7 F
18 ≥ 2 T	⁻14 ≤ ⁻15 F	7 ≤ 7 T
18 ≥ 18 T	⁻14 ≤ ⁻14 T	⁻7 ≤ 7 T
⁻3 ≤ ⁻3 T	0 > ⁻8 T	10 ≤ 24 T
⁻3 < ⁻3 F	0 ≥ ⁻8 T	13 > ⁻2 T
⁻3 < 0 T	⁻8 ≤ 0 T	⁻4 ≥ ⁻1 F

Book 5, Page 12

 is the international road sign for "Passing Permitted."

 is the sign for "No Passing."

Can you figure out what each sign below means?

 is the sign for No smoking.

 is the sign for No camping.

 is the sign for No picnicking.

 is the sign for No bicycling.

A slash is used on international road signs to mean "No." We use the same idea to make up new symbols in algebra. In these symbols the slash means "is not."

$7 \ne 4$ means "7 *is not equal to* 4."

$0 \not> 5$ means "0 *is not greater than* 5."

$2 \not< 2$ means "2 *is not less than* 2."

$10 \not\ge 12$ means "10 *is not greater than or equal to* 12."

$3 \not\le 0$ means "3 *is not less than or equal to* 0."

You write the meaning of each sentence below.

$8 \not< 4$ means "8 is not less than 4."

$0 \not> 3$ means "0 is not greater than 3."

$5 \not\ge 6$ means "5 is not greater than or equal to 6."

$^-10 \ne 10$ means "⁻10 is not equal to 10."

$x + 4 \ne 9$ means "x+4 is not equal to 9."

$^-2x \not< 10$ means "⁻2x is not less than 10."

$\frac{x}{3} \not\ge 5$ means "$\frac{x}{3}$ is not greater than or equal to 5."

Key to Algebra – ANSWERS

Book 5, Page 13

A number is a **solution** of an inequality if it makes the inequality true when you try it in place of x.

These numbers are solutions of $x + {}^-4 < 5$:			These numbers are *not* solutions of $x + {}^-4 < 5$:		
7	because	$7 + {}^-4 < 5$	10	because	$10 + {}^-4 \nless 5$
$^-1$	because	$^-1 + {}^-4 < 5$	9	because	$9 + {}^-4 \nless 5$
0	because	$0 + {}^-4 < 5$	25	because	$25 + {}^-4 \nless 5$

Try to find at least five integers which are solutions for each equation or inequality. If there aren't five integer solutions, list as many as you can find.

$x > {}^-3$	$^-2, {}^-1, 0, 1, 2, \ldots$	$x + 8 = 10$	2
$x \nless 0$	$0, 1, 2, 3, 4, \ldots$	$x + 8 < 10$	$1, 0, {}^-1, {}^-2, {}^-3, \ldots$
$x \neq 1$	$2, 3, 4, 5, 6, \ldots$ $0, {}^-1, {}^-2, {}^-3, {}^-4, \ldots$	$x + 8 > 10$	$3, 4, 5, 6, 7, \ldots$
$x \leq x$	Every integer is a solution.	$x + 8 \geq 10$	$2, 3, 4, 5, 6, \ldots$
$x < x$	No integer is a solution.	$x + 8 \leq 10$	$2, 1, 0, {}^-1, {}^-2, \ldots$
$x \neq x$	No integer is a solution.	$x + 8 \neq 10$	$3, 4, 5, 6, 7, \ldots$ $1, 0, {}^-1, {}^-2, {}^-3, \ldots$
$x = x$	Every integer is a solution.	$x + 8 \ngtr 10$	$2, 1, 0, {}^-1, {}^-2, \ldots$
$x > x$	No integer is a solution.	$x + 8 \nless 10$	$2, 3, 4, 5, 6, \ldots$
$x < x^2$	$^-1, {}^-2, {}^-3, {}^-4, {}^-5, \ldots$ $2, 3, 4, 5, 6, \ldots$	$x + 8 \nleq 10$	$3, 4, 5, 6, 7, \ldots$

Book 5, Page 14

Absolute Value

When we multiply or divide integers we get the *amount* of the answer by multiplying or dividing and the *sign* of the answer by following the rules for signs. The amount of a number is often called its **absolute value**. We put the symbol | | around a number when we want to talk about its absolute value.

$\lvert 6 \rvert = 6$	means	"The absolute value of 6 is 6."
$\lvert {}^-6 \rvert = 6$	means	"The absolute value of $^-6$ is 6."
$\lvert 0 \rvert = 0$	means	"The absolute value of 0 is 0."

Finding the absolute value of a number is easy. Just get rid of its sign. You find each absolute value below.

$\lvert {}^-4 \rvert = 4$	$\lvert {}^-19 \rvert = 19$	$\lvert {}^-\frac{2}{3} \rvert = \frac{2}{3}$	$\lvert 3 - 9 \rvert = \lvert {}^-6 \rvert = 6$
$\lvert 7 \rvert = 7$	$\lvert 19 \rvert = 19$	$\lvert {}^-1.3 \rvert = 1.3$	$\lvert 4({}^-5) \rvert = \lvert {}^-20 \rvert = 20$
$\lvert {}^-10 \rvert = 10$	$\lvert 0.19 \rvert = .19$	$\lvert 0.027 \rvert = 0.027$	$\lvert ({}^-6)({}^-4) \rvert = \lvert 24 \rvert = 24$

Use trial and error to find as many solutions as you can for each equation below.

(Both 5 and $^-5$ have absolute values of 5.)

$\lvert x \rvert = 5$	$5, {}^-5$	$\lvert x+2 \rvert = 4$	$2, {}^-6$
$\lvert x \rvert = 8$	$8, {}^-8$	$\lvert 3x \rvert = 27$	$9, {}^-9$
$\lvert x \rvert = 0$	0	$\lvert x - 5 \rvert = 2$	$7, 3$
$\lvert x \rvert = {}^-5$	no solution	$\lvert x - 8 \rvert = 0$	8

For each inequality below, try to find at least five integers which are solutions.

(7, 10 and 12 have absolute values greater than 6, but so do $^-7$, $^-10$ and $^-12$.)

$\lvert x \rvert > 6$	$7, 10, 12, {}^-7, {}^-10, {}^-12$	$\lvert x \rvert \nless 7$	$7, 8, 9, 10, 11, \ldots$ $^-7, {}^-8, {}^-9, {}^-10, {}^-11, \ldots$
$\lvert x \rvert < 6$	$^-5, {}^-4, {}^-3, {}^-2, {}^-1,$ $0, 1, 2, 3, 4, 5$	$\lvert x \rvert \ngtr 10$	(Every integer from $^-10$ to 10 is a solution.)
$\lvert x \rvert \geq 0$	(Every integer is a solution.)	$\lvert x \rvert > x$	(Every negative integer is a solution.)
$\lvert x \rvert \neq 3$	(Every integer except 3 and $^-3$ is a solution.)	$\lvert x \rvert \leq 2$	$^-2, {}^-1, 0, 1, 2$

Book 5, Page 15

Graphing Inequalities

Look at the last equation on the previous page. The five integers which are solutions are $^-2$, $^-1$, 0, 1 and 2. We could easily make a graph of this set of solutions.

$\lvert x \rvert \leq 2$

Some inequalities, like $x > {}^-3$, have an infinite number of solutions. It would be impossible to list all the integers which are solutions, but we could show the solution set by *starting* a list and then using three dots to show that it continues on and on.

$x > {}^-3$ $\{{}^-2, {}^-1, 0, 1, 2, 3, \ldots\}$

We could also graph the set of solutions using a darkened arrow on the right to show that the dots continue to the right.

$x > {}^-3$ $\{{}^-2, {}^-1, 0, 1, 2, 3, \ldots\}$

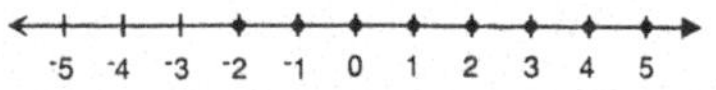

For each inequality, show the integers which are solutions in two ways: by making a list and by graphing.

	List
$x < 1$	$\{0, {}^-1, {}^-2, {}^-3, \ldots\}$
$x > 5$	$\{6, 7, 8, 9, \ldots\}$
$x \leq {}^-3$	$\{{}^-3, {}^-4, {}^-5, {}^-6, \ldots\}$
$x \geq {}^-4$	$\{{}^-4, {}^-3, {}^-2, {}^-1, \ldots\}$
$x > 210$	$\{211, 212, 213, \ldots\}$
$x \leq 0$	$\{0, {}^-1, {}^-2, {}^-3, \ldots\}$
$\lvert x \rvert < 4$	$\{3, 2, 1, 0, {}^-1, {}^-2, {}^-3\}$
$\lvert x \rvert \geq 2$	$\{2, 3, 4, 5, \ldots$ $^-2, {}^-3, {}^-4, {}^-5, \ldots\}$
$x + 1 \geq 5$	$\{4, 5, 6, 7, \ldots\}$

Graph

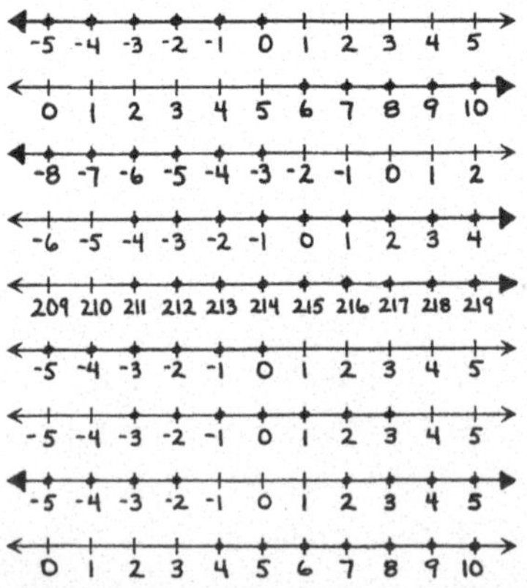

Book 5, Page 16

It would be impossible to list all the rational numbers which are solutions of the inequality $x > {}^-3$. $^-3$ is not a solution, but every rational number greater than $^-3$ is. We can show the **solution set** very clearly by graphing.

For each equation or inequality below, graph the set of all rational numbers which are solutions.

$x > 1$

$x < 2$

$x \leq 2$

$x \geq {}^-4$

$x \neq {}^-3$

$x \geq 1.5$

$x < 2\frac{1}{3}$

$x \neq 4.4$

$x \leq 0$

$x \ngtr 2$

$\lvert x \rvert > 3$

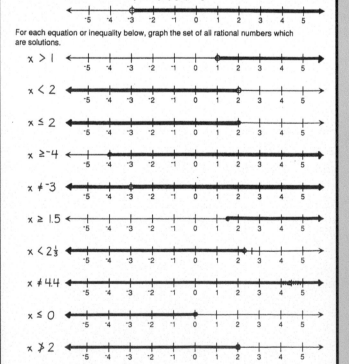

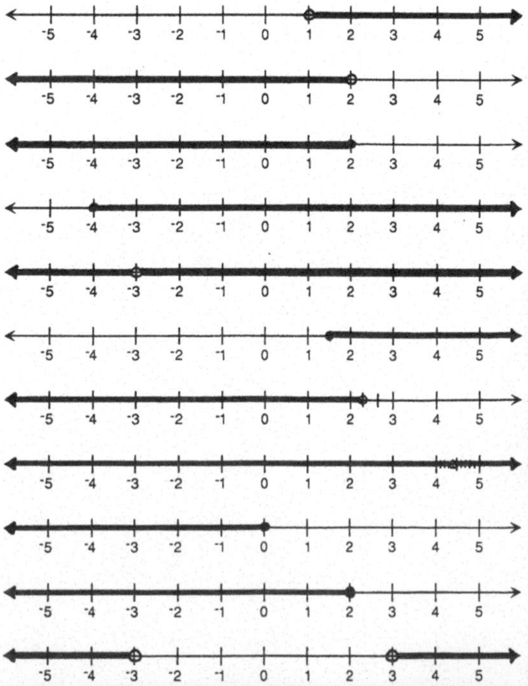

Key to Algebra – ANSWERS

Book 5, Page 17

The graph of an inequality depends on what kinds of numbers we allow as solutions. The set of numbers we allow as solutions is called the **replacement set**.

In the first problem below we have shown what the graph of $x < 4$ looks like when only integers are allowed as solutions and what it looks like when all rational numbers are allowed as solutions. Make a graph of each inequality for each replacement set.

Integers — *Rational Numbers*

$x < 4$

$x \le 0$

$x > 2.5$

$x \ge \frac{1}{3}$

$x < 3.5$

$|x| < 2$

$|x| < 2.5$

$|x| > 4$

$|x| > 0$

$x \ne {}^{-}4$

$x \ngtr 1$

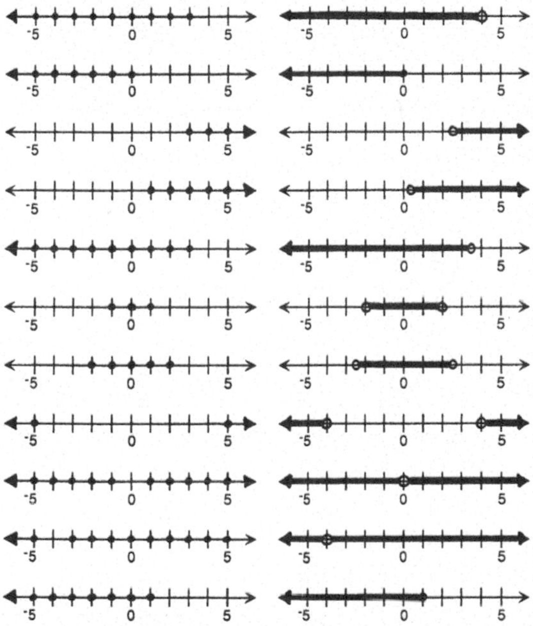

From now on in this book we'll always use rational numbers as our replacement set.

Book 5, Page 18

Solving Inequalities

We can tell if a number is a solution of an equation or inequality by substituting that number for the variable and seeing whether the result is true or false.

Is 8 a solution of $2x + 5 < x - 15$? $\quad 2x + 5 < x - 15$

We can substitute 8 for x to find out: $\quad 2\cdot 8 + 5 < 8 - 15$

$16 + 5 < 8 + {}^{-}15$

$21 < {}^{-}7$

So 8 is not a solution. false

Luckily we do not have to substitute every time we want to check a possible solution. We can use the Addition Principle to help us solve an inequality just as we used it to solve equations. The Addition Principle can help us find a simpler inequality with the same solution set. This is called **solving** the inequality. Look at this example:

$$2x + 5 < x - 15$$
$$x + 5 < {}^{-}15$$
$$x < {}^{-}20$$

Only numbers less than ${}^{-}20$ are solutions, so 8 could not be a solution. Neither could any other positive number.

Solve each inequality below using the Addition Principle.

$x + 19 \ge 3$ $x \ge {}^{-}16$	$x - 4 < 12$ $x < 16$	$x + 9 \le {}^{-}6$ $x \le {}^{-}15$
${}^{-}2 + x < 6$ $x < 8$	$11 + x \ge 1$ $x \ge {}^{-}10$	$x - 5 < {}^{-}3$ $x < 2$
$5 > x + 4$ $1 > x$ $x < 1$ (If 1 is bigger than x, then x is less than 1.)	$10 < x + 6$ $4 < x$ $x > 4$	$12 \ge x - 7$ $19 \ge x$ $x \le 19$
$5x + 1 > 4x + 15$ $x + 1 > 15$ $x > 14$	$7x - 9 < 6x + 11$ $x - 9 < 11$ $x < 20$	$x - 5 \le 2x + 8$ ${}^{-}5 \le x + 8$ ${}^{-}13 \le x$ $x \ge {}^{-}13$

Book 5, Page 19

Solve each inequality below using the Addition Principle. Draw a graph of each solution set.

$x - 7 > 0$ $x > 7$	$x - 9 < {}^{-}9$ $x < 0$	$6 \le x + 12$ ${}^{-}6 \le x$ $x \ge {}^{-}6$

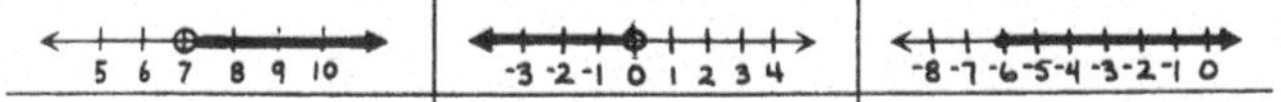

$2x + 1 > x - 4$ $x + 1 > {}^{-}4$ $x > {}^{-}5$	$x + 16 < 2x + 20$ $16 < x + 20$ ${}^{-}4 < x$ $x > {}^{-}4$	$5x - 6 \le 4x + 1$ $x - 6 \le 1$ $x \le 7$

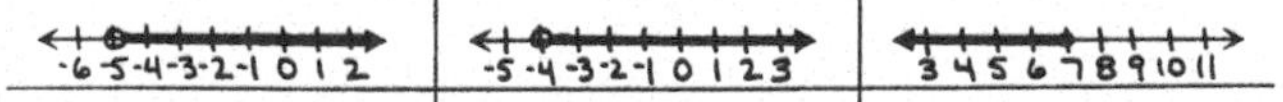

$3(x - 3) > 2x$ $3x - 9 > 2x$ $x - 9 > 0$ $x > 9$	$x - 4 + 11 \le 26$ $x + 7 \le 26$ $x \le 19$	$2(x - 1) < 3(x + 4)$ $2x - 2 < 3x + 12$ ${}^{-}2 < x + 12$ ${}^{-}14 < x$ $x > {}^{-}14$

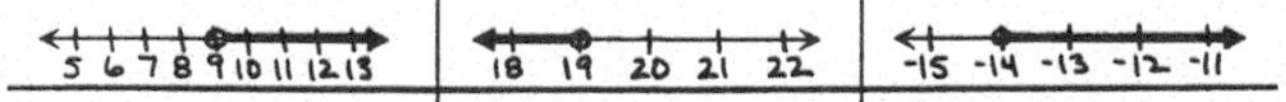

$x^2 - 5 < x^2 + x + 1$ ${}^{-}5 < x + 1$ ${}^{-}6 < x$ $x > {}^{-}6$	$x^2 + x + 8 \ge x^2 - 4$ $x + 8 \ge {}^{-}4$ $x \ge {}^{-}12$	$2x^2 + 22 > 2x^2 + x - 9$ $22 > x - 9$ $31 > x$ $x < 31$

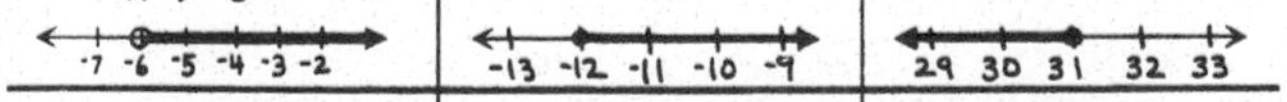

$(x + 2)(x - 2) < x^2 + x$ $x^2 - 4 < x^2 + x$ ${}^{-}4 < x$ $x > {}^{-}4$	$(x + 2)(x - 1) \ge x^2$ $x^2 - 1x + 2x - 2 \ge x^2$ $x^2 + x - 2 \ge x^2$ $x - 2 \ge 0$ $x \ge 2$	$(x - 4)(x + 5) < x^2 - 19$ $x^2 + 5x - 4x - 20 < x^2 - 19$ $x^2 + x - 20 < x^2 - 19$ $x - 20 < {}^{-}19$ $x < 1$

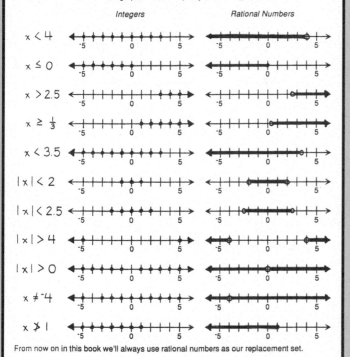

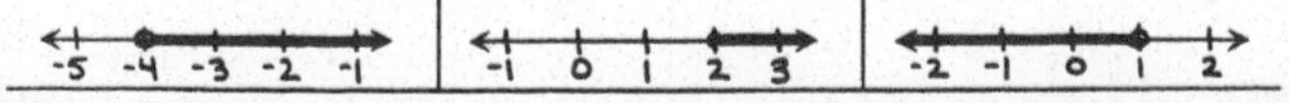

Book 5, Page 20

Maybe you are wondering whether there are Multiplication and Division Principles for Inequalities. The answer is yes, but they aren't quite the same as the Multiplication and Division Principles for Equations. When we multiply or divide both sides of an inequality by a *positive* number, we do get an equivalent inequality. But when we multiply or divide both sides by a *negative* number, we must *reverse the inequality sign* to get an equivalent inequality. Look at these sentences to see why:

Multiplying by 2:	${}^{-}6 < 10$ ← true ${}^{-}6\cdot 2 < 10\cdot 2$ ← true ${}^{-}12 < 20$	Dividing by 2:	${}^{-}6 < 10$ ← true $\frac{-6}{2} < \frac{10}{2}$ ← true ${}^{-}3 < 5$
Multiplying by ${}^{-}2$:	${}^{-}6 < 10$ ← true ${}^{-}6\cdot{}^{-}2 < 10\cdot{}^{-}2$ ← false $12 < {}^{-}20$ $12 > {}^{-}20$ ← true with < switched to >	Dividing by ${}^{-}2$:	${}^{-}6 < 10$ ← true $\frac{-6}{-2} < \frac{10}{-2}$ ← false $3 < {}^{-}5$ $3 > {}^{-}5$ ← true with < switched to >

Solve each inequality using the Multiplication and Division Principles. Remember to switch the direction of the inequality sign if you multiply or divide by a negative number.

(5 is positive. Leave the ≥ sign alone.) $\frac{5x}{5} \ge \frac{-30}{5}$ $x \ge {}^{-}6$	(${}^{-}4$ is negative. Switch < to >.) $\frac{x}{-4} < 6$ ${}^{-}4\cdot(\frac{x}{-4}) > (6)\cdot{}^{-}4$ $x > {}^{-}24$	$5\cdot\frac{x}{5} < 7\cdot 5$ $x < 35$
${}^{-}9x \le 72$ $\frac{-9x}{-9} \ge \frac{72}{-9}$ $x \ge {}^{-}8$	$\frac{x}{-6} < 8$ ${}^{-}6\cdot\frac{x}{-6} > 8\cdot{}^{-}6$ $x > {}^{-}48$	$3\cdot\frac{x}{3} \ge {}^{-}4\cdot 3$ $x \ge {}^{-}12$
$\frac{x}{-10} \ge {}^{-}3$ ${}^{-}10(\frac{x}{-10}) \le ({}^{-}3)({}^{-}10)$ $x \le 30$	(${}^{-}x = {}^{-}1x$) ${}^{-}x > 4$ $({}^{-}1)({}^{-}1x) < (4)({}^{-}1)$ $x < {}^{-}4$	$\frac{11x}{11} < \frac{-33}{11}$ $x < {}^{-}3$
$\frac{x}{-5} \le 12$ ${}^{-}5(\frac{x}{-5}) \ge (12)({}^{-}5)$ $x \ge {}^{-}60$	$\frac{2x}{2} > \frac{-5}{2}$ $x > \frac{-5}{2}$	${}^{-}3x < 0$ $\frac{-3x}{-3} > \frac{0}{-3}$ $x > 0$

Key to Algebra – ANSWERS

Book 5, Page 21

Solving each inequality below takes more than one step. Remember to switch the inequality sign whenever you multiply or divide both sides by a negative number.

(Adding ⁻5 to each side is O.K.)

$^-3x + 5 > 22$
$^-3x > 17$
(Dividing by ⁻3 means I have to reverse >.)
$\frac{^-3x}{^-3} < \frac{17}{^-3}$
$x < \frac{^-17}{3}$

$5x + 1 \le {^-44}$
$\frac{5x}{5} \le \frac{^-45}{5}$
$x \le {^-9}$

$^-2x + 15 < 7$
$^-2x < {^-8}$
$\frac{^-2x}{^-2} > \frac{^-8}{^-2}$
$x > 4$

$2 \cdot \frac{x+15}{2} < {^-5} \cdot 2$
$x + 15 < {^-10}$
$x < {^-25}$

$\frac{x+9}{^-3} > 3$
$^-3 \cdot \frac{x+9}{^-3} < 3 \cdot {^-3}$
$x + 9 < {^-9}$
$x < {^-18}$

$\frac{x-6}{^-10} \ge {^-4}$
$^-10 \cdot \frac{x-6}{^-10} \le {^-4} \cdot {^-10}$
$x - 6 \le 40$
$x \le 46$

$4 \cdot \frac{^-5x}{4} \ge 10 \cdot 4$
$^-5x \ge 40$
$\frac{^-5x}{^-5} \le \frac{40}{^-5}$
$x \le {^-8}$

$7 \cdot \frac{3x}{7} < 1 \cdot 7$
$\frac{3x}{3} < \frac{7}{3}$
$x < \frac{7}{3}$

$3 \cdot \frac{^-4x}{3} > {^-12} \cdot 3$
$^-4x > {^-36}$
$\frac{^-4x}{^-4} < \frac{^-36}{^-4}$
$x < 9$

$\frac{x}{6} + 14 > 9$
$6 \cdot \frac{x}{6} > {^-5} \cdot 6$
$x > {^-30}$

$\frac{x}{^-7} + 5 \le 12$
$\frac{x}{^-7} \le 7$
$^-7 \cdot \frac{x}{^-7} \ge 7 \cdot {^-7}$
$x \ge {^-49}$

$\frac{x}{2} - 10 < {^-19}$
$2 \cdot \frac{x}{2} < {^-9} \cdot 2$
$x < {^-18}$

$2x - 7x \le 3$
$^-5x \le 3$
$\frac{^-5x}{^-5} \ge \frac{3}{^-5}$
$x \ge \frac{^-3}{5}$

$^-3(x-5) > 21$
$^-3x + 15 > 21$
$^-3x > 6$
$\frac{^-3x}{^-3} < \frac{6}{^-3}$
$x < {^-2}$

$9 - x \le 7$
$^-x \le {^-2}$
$^-1x \le {^-2}$
$\frac{^-1x}{^-1} \ge \frac{^-2}{^-1}$
$x \ge 2$

Book 5, Page 22

Here's how Sandy and Terry solved the last inequality on page 21.

Sandy

$9 - x \le 7$
$9 \le 7 + x$
$2 \le x$
$x \ge 2$

Terry

$9 - x \le 7$
$^-x \le {^-2}$
$^-1x \le {^-2}$
$\frac{^-1x}{^-1} \ge \frac{^-2}{^-1}$
$x \ge 2$

Both Sandy and Terry ended up with the same solution set. Whose method do you like better? Why?

Both methods work. Use either one to solve each inequality below.

$2 - x > 16$
$2 > 16 + x$
$^-14 > x$
$x < {^-14}$

$25 \le 10 - x$
$x + 25 \le 10$
$x \le {^-15}$

$^-5x + 9 \ge 3x$
$\frac{9}{8} \ge \frac{8x}{8}$
$\frac{9}{8} \ge x$
$x \le \frac{9}{8}$

$3x + 15 > 3 + 7x$
$15 > 3 + 4x$
$\frac{12}{4} > \frac{4x}{4}$
$3 > x$
$x < 3$

$x + 20 < 5x - 8$
$20 < 4x - 8$
$\frac{28}{4} < \frac{4x}{4}$
$7 < x$
$x > 7$

$18 + x \ge 12 + 7x$
$18 - 6x \ge 12$
$^-6x \ge {^-6}$
$\frac{^-6x}{^-6} \le \frac{^-6}{^-6}$
$x \le 1$

$5(3 - x) < 50$
$15 - 5x < 50$
$^-5x < 35$
$\frac{^-5x}{^-5} > \frac{35}{^-5}$
$x > {^-7}$

$2 \cdot \frac{4-x}{2} > 12 \cdot 2$
$4 - x > 24$
$^-x > 20$
$x < {^-20}$

$12 - 7x \le 12$
$^-7x \le 0$
$\frac{^-7x}{^-7} \ge \frac{0}{^-7}$
$x \ge 0$

Book 5, Page 23

Solve each inequality and graph the solution set.

$3x - 4x > 6$
$^-1x > 6$
$\frac{^-1x}{^-1} < \frac{6}{^-1}$
$x < {^-6}$

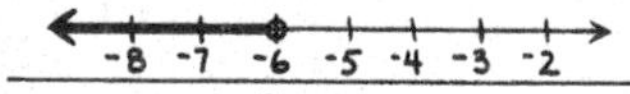

$x + 12 \ge 2x - 5$
$12 \ge x - 5$
$17 \ge x$
$x \le 17$

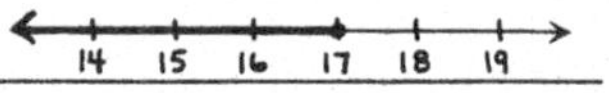

$2 \cdot \frac{x-7}{2} < {^-6} \cdot 2$
$x - 7 < {^-12}$
$x < {^-5}$

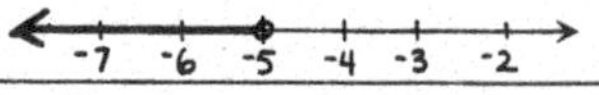

$\frac{x}{2} - 7 < {^-6}$
$2 \cdot \frac{x}{2} < 1 \cdot 2$
$x < 2$

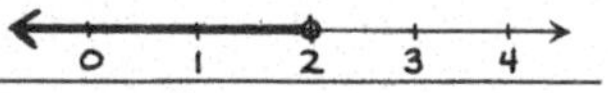

$7x - 2x - x \ge 24 + 3x$
$4x \ge 24 + 3x$
$x \ge 24$

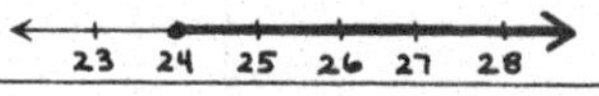

$\frac{4x}{3} + 1 < 4$
$3 \cdot \frac{4x}{3} < 3 \cdot 3$
$\frac{4x}{4} < \frac{9}{4}$
$x < 2\frac{1}{4}$

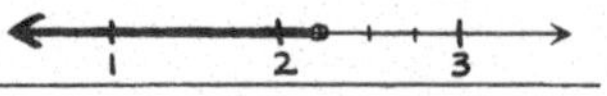

$^-16 + 4x > 10 - x$
$^-16 + 5x > 10$
$\frac{5x}{5} > \frac{26}{5}$
$x > 5\frac{1}{5}$

$3(x-4) - 9x \ge 2x - 4$
$3x - 12 - 9x \ge 2x - 4$
$^-6x - 12 \ge 2x - 4$
$^-8x - 12 \ge {^-4}$
$^-8x \ge 8$
$\frac{^-8x}{^-8} \le \frac{8}{^-8}$
$x \le {^-1}$

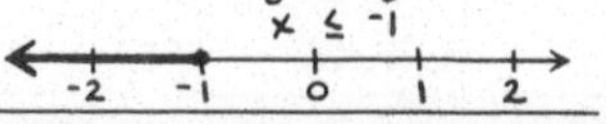

Book 5, Page 24

When an absolute value sign appears in an equation or inequality, you should *not* use the Addition, Multiplication or Division Principle to simplify an expression *inside* the absolute value sign. Instead, think about what the absolute value sign means.

(The number inside the | | sign could be either 6 or ⁻6.)

$|x + 1| = 6$

$x + 1 = 6$ or $x + 1 = {^-6}$

$x = 5$ or $x = {^-7}$

To make sure that both 5 and ⁻7 are solutions, we can substitute each for x.

Check: $|5 + 1| = |6| = 6$
$|{^-7} + 1| = |{^-6}| = 6$

Solve each equation, and check your solutions.

$|x + 5| = 7$
$x + 5 = 7$ or $x + 5 = {^-7}$
$x = 2$ or $x = {^-12}$
Check: $|2 + 5| = |7| = 7$
$|{^-12} + 5| = |{^-7}| = 7$

$|x - 10| = 2$
$x - 10 = 2$ or $x - 10 = {^-2}$
$x = 12$ or $x = 8$
Check: $|12 - 10| = |2| = 2$
$|8 - 10| = |{^-2}| = 2$

$\left|\frac{x}{5}\right| = 2$
$5 \cdot \frac{x}{5} = 2 \cdot 5$ or $5 \cdot \frac{x}{5} = {^-2} \cdot 5$
$x = 10$ or $x = {^-10}$
Check: $\left|\frac{10}{5}\right| = |2| = 2$
$\left|\frac{^-10}{5}\right| = |{^-2}| = 2$

$|{^-3x}| = 4$
$\frac{^-3x}{^-3} = \frac{4}{^-3}$ or $\frac{^-3x}{^-3} = \frac{^-4}{^-3}$
$x = \frac{^-4}{3}$ or $x = \frac{4}{3}$
Check: $\left|{^-3}\left(\frac{^-4}{3}\right)\right| = |4| = 4$
$\left|{^-3}\left(\frac{4}{3}\right)\right| = |{^-4}| = 4$

$|3x + 4| = 10$
$3x + 4 = 10$ or $3x + 4 = {^-10}$
$3x = 6$ $3x = {^-14}$
$x = 2$ $x = \frac{^-14}{3}$
Check: $|3 \cdot 2 + 4| = |6 + 4| = |10| = 10$
$\left|3\left(\frac{^-14}{3}\right) + 4\right| = |{^-14} + 4| = |{^-10}| = 10$

$|12x - 6| = 6$
$12x - 6 = 6$ or $12x - 6 = {^-6}$
$12x = 12$ $12x = 0$
$x = 1$ $x = 0$
Check: $|12 \cdot 1 - 6| = |12 - 6| = |6| = 6$
$|12 \cdot 0 - 6| = |0 - 6| = |{^-6}| = {^-6}$

Key to Algebra – ANSWERS

Book 5, Page 25

Solve each equation.

(The 3 is outside the | | sign, so I can use the Addition Principle.)

$|x| + 3 - 3 = 20 - 3$
$|x| = 17$
$x = 17$ or -17

$|x| - 8 = -2$
$|x| = 6$
$x = 6$ or -6

$|x| + 8 = 11$
$|x| = 3$
$x = 3$ or -3

$2|x| - 5 = 3$
$\frac{2|x|}{2} = \frac{8}{2}$
$|x| = 4$
$x = 4$ or -4

$|4x - 5| = 7$
$4x - 5 = 7$ or $4x - 5 = -7$
$4x = 12$ $\qquad \frac{4x}{4} = \frac{-2}{4}$
$x = 3$ $\qquad x = \frac{-2}{4}$

$4|x| - 5 = 7$
$\frac{4|x|}{4} = \frac{12}{4}$
$|x| = 3$
$x = 3$ or -3

$|x + 2| - 3 = 7$
$|x + 2| = 10$
$x + 2 = 10$ or $x + 2 = -10$
$x = 8$ or $x = -12$

$|x + 5| + 4 = 12$
$|x + 5| = 8$
$x + 5 = 8$ or $x + 5 = -8$
$x = 3$ or $x = -13$

$|2x - 3| + 2 = 11$
$|2x - 3| = 9$
$2x - 3 = 9$ or $2x - 3 = -9$
$2x = 12$ $\qquad 2x = -6$
$x = 6$ $\qquad x = -3$

$-3|x| + 5 = -4$
$\frac{-3|x|}{-3} = \frac{-9}{-3}$
$|x| = 3$
$x = 3$ or -3

Book 5, Page 26

Solve and check each inequality.

You can't check all the solutions, so pick a few samples.

$|x| + 2 < 11$
(Numbers from 9 down to -9 will work.)
$|x| < 9$
$x < 9$ and $x > -9$
Check:
$6: |6| + 2 = 6 + 2 = 8 < 11$
$-4: |-4| + 2 = 4 + 2 = 6 < 11$

$|x - 2| > 3$
(What's inside must be over 3 or under -3.)
$x - 2 > 3$ or $x - 2 < -3$
$x > 5$ or $x < -1$
Check:
$9: |9 - 2| = 7 > 3$
$-4: |-4 - 2| = |-6| = 6 > 3$

$|x| > 12$
$x > 12$ or $x < -12$
Check: $15: |15| = 15 > 12$
$-20: |-20| = 20 > 12$

$|x| \le 4$
$x \le 4$ and $x \ge -4$
Check: $4: |4| = 4 \le 4$
$1: |1| = 1 \le 4$
$-3: |-3| = 3 \le 4$

$|x| + 4 - 4 < 20 - 4$
$|x| < 16$
$x < 16$ and $x > -16$
Check: $10: |10| + 4 = 10 + 4 = 14 < 20$
$-15: |-15| + 4 = 15 + 4 = 19 < 20$

$|x| - 6 + 6 > 1 + 6$
$|x| > 7$
$x > 7$ or $x < -7$
Check: $10: |10| - 6 = 10 - 6 = 4 > 1$
$-8: |-8| - 6 = 8 - 6 = 2 > 1$

$|x - 1| < 5$
$x - 1 < 5$ and $x - 1 > -5$
$x < 6$ and $x > -4$
Check: $5: |5 - 1| = |4| = 4 < 5$
$-3: |-3 - 1| = |-4| = 4 < 5$

$|x + 10| > 5$
$x + 10 > 5$ or $x + 10 < -5$
$x > -5$ or $x < -15$
Check: $-3: |-3 + 10| = |7| = 7 > 5$
$-20: |-20 + 10| = |-10| = 10 > 5$

$|12x| \ge 24$
$\frac{12x}{12} \ge \frac{24}{12}$ or $12x \le -24$
$x \ge 2$ or $x \le -2$
Check:
$2: |12 \cdot 2| = |24| = 24 \ge 24$
$5: |12 \cdot 5| = |60| = 60 \ge 24$
$-2: |12 \cdot -2| = |-24| = 24 \ge 24$
$-10: |12 \cdot -10| = |-120| = 120 \ge 24$

$\left|\frac{x}{3}\right| < 8$
$3 \cdot \frac{x}{3} < 8 \cdot 3$ and $3 \cdot \frac{x}{3} > -8 \cdot 3$
$x < 24$ and $x > -24$
Check: $21: \left|\frac{21}{3}\right| = |7| = 7 < 8$
$-21: \left|\frac{-21}{3}\right| = |-7| = 7 < 8$

Book 5, Page 27

Relations

In algebra, =, <, >, ≤ and ≥ are sometimes called **relations**. Thinking of relations in your family can help you understand relations in algebra. Relations in your family are people connected to you in certain ways. You have different kinds of relations: mother, father, brothers, sisters, aunts, uncles, etc. Look at this family tree:

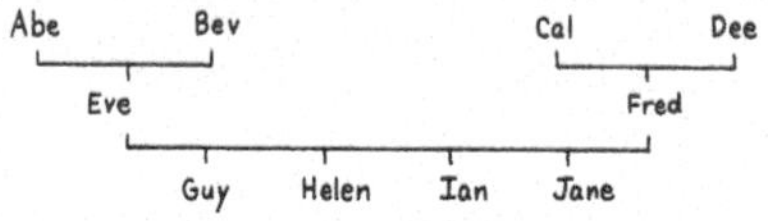

The tree shows Abe is married to Bev, and Eve is their child. Cal is married to Dee, and their child is Fred. Guy, Helen, Ian and Jane are the children of Eve and Fred.

Let's see how some of the people in this family are related:

Sisters: We will call the sister relation S. $S(x)$ means "a sister of x."

S(Guy) = Helen means "A sister of Guy is Helen."
S(Guy) = Jane means "A sister of Guy is Jane."
S(Ian) = Helen means "A sister of Ian is Helen."
S(Ian) = Jane means "A sister of Ian is Jane."
S(Jane) = Helen means "A sister of Jane is Helen."
S(Helen) = Jane means "A sister of Helen is Jane."

Brothers: We will call this relation B. $B(x)$ means "a brother of x."

B(Ian) = Guy	B(Jane) = Guy	B(Jane) = Ian
B(Guy) = Ian	B(Helen) = Guy	B(Helen) = Ian

Fathers: We will call this relation F. $F(x)$ means "the father of x."

F(Eve) = Abe	F(Ian) = Fred	F(Guy) = Fred
F(Fred) = Cal	F(Helen) = Fred	F(Jane) = Fred

Grandmothers: We will call this relation G. $G(x)$ means "a grandmother of x."

G(Helen) = Bev	G(Jane) = Bev	G(Ian) = Bev
G(Helen) = Dee	G(Jane) = Dee	G(Ian) = Dee
G(Guy) = Bev	G(Guy) = Dee	

Book 5, Page 28

Relations in families pair people with other people. Relations in algebra pair numbers with other numbers. Here are some relations which involve numbers:

The "greater than" relation: We will use $G(x)$ to mean "a number greater than x."

G(5) = 7 means "A number greater than 5 is 7." or $7 > 5$

G(5) = 10.4 means "A number greater than 5 is 10.4." or $10.4 > 5$

Name some other numbers which can be paired up with 5 in this relation:

G(5) = 6 $\qquad$ G(5) = 8 $\qquad$ G(5) = 13.9 $\qquad$ G(5) = 100

Find a number to make each of these true:

G(2) = 3 $\qquad$ G(-6) = 0 $\qquad$ G(100) = 200 $\qquad$ G(4.8) = 4.9

The "less than" relation: We will use $L(x)$ to mean "a number less than x."

L(3) = 2 means "A number less than 3 is 2." or $2 < 3$

L(3) = $-1\frac{1}{3}$ means "A number less than 3 is $-1\frac{1}{3}$." or $-1\frac{1}{3} < 3$

L(-1) = means "A number less than -1 is -2." or $-2 < -1$

L(-2.5) = means "A number less than -2.5 is -3." or $-3 < -2.5$

The "equality" relation: We will use $E(x)$ to mean "a number equal to x."

E(2) = 2 means "A number equal to 2 is 2." or $2 = 2$

E(-6) = means "A number equal to -6 is -6." or $-6 = -6$

E(11.2) = means "A number equal to 11.2 is 11.2." or $11.2 = 11.2$

The "less than or equal to" relation: We will use $T(x)$ to mean "a number less than or equal to x."

T(4) = 0 means "A number less than or equal to 4 is 0." or $0 \le 4$

$T\left(\frac{3}{5}\right) = \frac{3}{5}$ means "A number less than or equal to $\frac{3}{5}$ is $\frac{3}{5}$." or $\frac{3}{5} \le \frac{3}{5}$

T(-6.3) = means "A number less than or equal to -6.3 is -7." or $-7 \le -6.3$

Key to Algebra – ANSWERS

Book 5, Page 29

Functions

Some people like to think of relations as "input-output" machines.

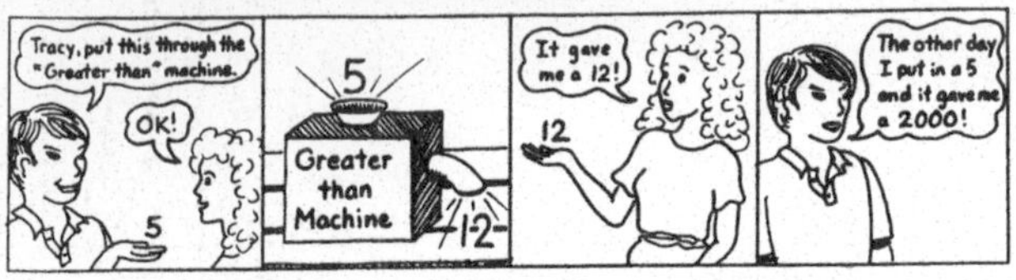

As you can see, the output of a "greater than" machine can't be predicted.
All we know is that it will put out a number greater than the number which is put in.
Here are some possibilities:

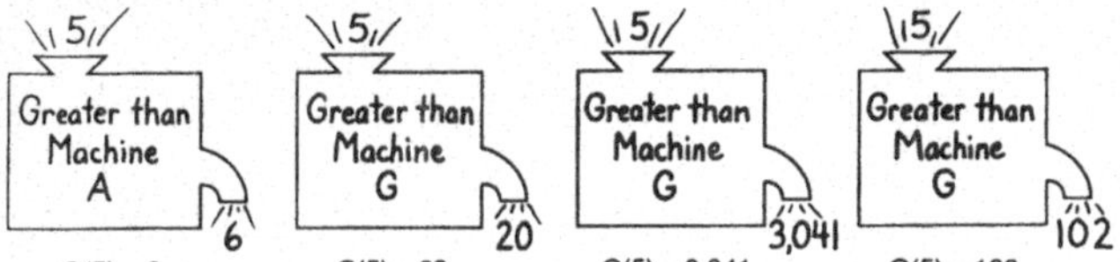

G(5) = 6 G(5) = 20 G(5) = 3,041 G(5) = 102

Some kinds of relation machines are completely predictable.

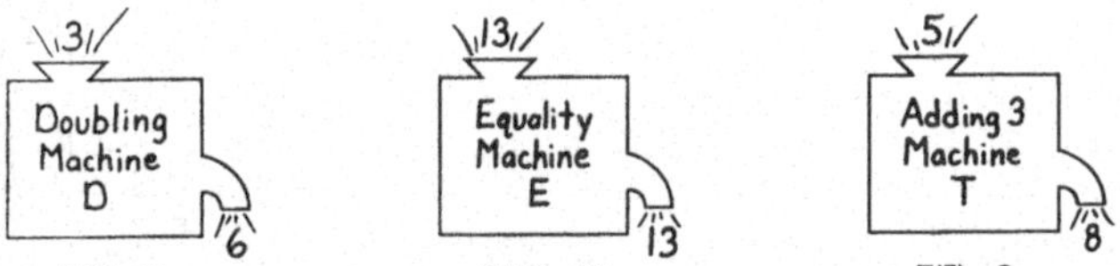

D(3) = 6 E(13) = 13 T(5) = 8

These completely predictable relations are called **functions**. A function has only one possible output for each input.

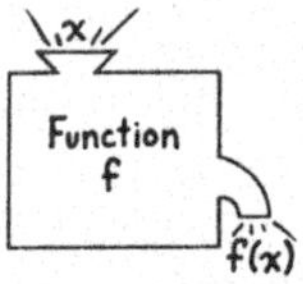

Book 5, Page 30

Here are some more examples of functions.

The "absolute value" function: $A(x)$ will mean "the absolute value of x." $A(x) = |x|$

$A(7) = \|7\| = 7$	$A(^-5.3) = \|^-5.3\| = 5.3$
$A(^-7) = \|^-7\| = 7$	$A(4.71) = \|4.71\| = 4.71$
$A(3) = \|3\| = 3$	$A(\frac{1}{2}) = \|\frac{1}{2}\| = \frac{1}{2}$
$A(^-3) = \|^-3\| = 3$	$A(\frac{-3}{4}) = \|\frac{-3}{4}\| = \frac{3}{4}$
$A(^-12) = \|^-12\| = 12$	$A(^-6.6) = \|^-6.6\| = 6.6$

The "next integer" function: $N(x)$ will mean "the integer after x." $N(x) = x + 1$

$N(10) = 11$	$N(^-16) = ^-15$
$N(9) = 10$	$N(^-101) = ^-100$
$N(0) = 1$	$N(25) = 26$

The "opposite" function: $P(x)$ will mean "the opposite of x." $P(x) = -x$

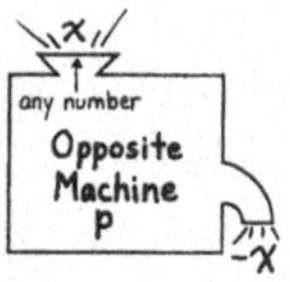

$P(2) = ^-2$	$P(\frac{3}{5}) = \frac{-3}{5}$	$P(3.5) = ^-3.5$
$P(^-1) = 1$	$P(\frac{-1}{2}) = \frac{1}{2}$	$P(0.6) = ^-0.6$
$P(0) = 0$	$P(\frac{-10}{7}) = \frac{10}{7}$	$P(^-0.003) = 0.003$

The "squaring" function: $S(x)$ will mean "the square of x." $S(x) = x^2$

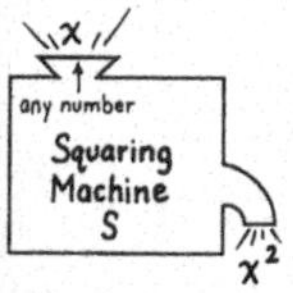

$S(3) = 9$	$S(^-3) = 9$	$S(12) = 144$
$S(4) = 16$	$S(^-4) = 16$	$S(100) = 10000$
$S(5) = 25$	$S(^-5) = 25$	$S(^-100) = 10000$
$S(6) = 36$	$S(^-6) = 36$	$S(0.5) = .25$

Book 5, Page 31

Here are some functions we use in everyday life:

The "first class postage" function: $P(x)$ is the first class postage on a letter weighing x ounces. It costs 25¢ to mail a letter weighing one ounce or less. For each additional ounce or part of an ounce you pay 20¢ more.

$P(1) =$ 25¢	$P(\frac{1}{4}) =$ 25¢	$P(0.7) =$ 25¢
$P(2) =$ 45¢	$P(\frac{3}{4}) =$ 25¢	$P(7) =$ \$1.45
$P(3) =$ 65¢	$P(1\frac{1}{4}) =$ 45¢	$P(7.3) =$ \$1.65
$P(4) =$ 85¢	$P(2\frac{1}{4}) =$ 65¢	$P(3.7) =$ 85¢
$P(5) =$ \$1.05	$P(3\frac{1}{2}) =$ 85¢	$P(4.9) =$ \$1.05

The "feet-to-inches" function: $I(x)$ is the number of inches in x feet.
There are 12 inches in one foot, so $I(1) = 12$.

$I(2) = 24$	$I(5) = 60$	$I(\frac{1}{2}) = 6$
$I(3) = 36$	$I(9) = 108$	$I(1\frac{1}{2}) = 18$
$I(4) = 48$	$I(15) = 180$	$I(\frac{1}{3}) = 4$

The "sales tax" function: $S(x)$ is the sales tax on a taxable purchase of x dollars.
To figure these out, you need to know the sales tax percentage for your state. Try to get a copy of the sales tax table which many stores use.

$S(.10) =$ Answers	$S(.50) =$ will	$S(.87) =$ vary
$S(.20) =$	$S(1.00) =$	$S(1.10) =$
$S(.30) =$	$S(2.00) =$	$S(7.75) =$

The "days-in-a-year" function: $D(x)$ is the number of days in the year x.
Leap years have 366 days. Others have 365 days.

$D(1980) = 366$	$D(1971) = 365$	$D(1776) = 366$
$D(1950) = 365$	$D(1900) = 365$	$D(2000) = 366$

Book 5, Page 32

Instead of describing a function in words, we often define it by giving an algebraic expression for the number which is paired up with x. You finish this example:

The "f" function: $f(x)$ is one more than 3 times x. $f(x) = 3x + 1$

$f(5) = 3(5) + 1 = 15 + 1 = 16$ $f(8) =$

$f(^-2) = 3(^-2) + 1 = ^-6 + 1 = ^-5$ $f(^-8) =$

Each function below is defined by an algebraic expression. Find the numbers asked for by substituting in the expression.

$g(x) = 3x - 1$ $g(^-4) = 3(^-4) - 1 = ^-12 - 1 = ^-13$ $g(7) = 3(7) - 1 = 21 - 1 = 20$ $g(5) = 3(5) - 1 = 15 - 1 = 14$	$h(x) = 3(x - 1)$ $h(^-4) = 3(^-4 - 1) = 3(^-5) = ^-15$ $h(7) = 3(7 - 1) = 3(6) = 18$ $h(5) = 3(5 - 1) = 3(4) = 12$
$p(x) = x^2 + 4$ $p(3) = 3^2 + 4 = 9 + 4 = 13$ $p(1) = 1^2 + 4 = 1 + 4 = 5$ $p(^-10) = ^-10^2 + 4 = 100 + 4 = 104$	$q(x) = (x + 4)^2$ $q(3) = (3 + 4)^2 = 7^2 = 49$ $q(1) = (1 + 4)^2 = 5^2 = 25$ $q(^-10) = (^-10 + 4)^2 = ^-6^2 = 36$
$r(x) = \|x + 6\|$ $r(3) = \|3 + 6\| = \|9\| = 9$ $r(^-10) = \|^-10 + 6\| = \|^-4\| = 4$ $r(^-6) = \|^-6 + 6\| = \|0\| = 0$	$s(x) = \|x\| + 6$ $s(3) = \|3\| + 6 = 3 + 6 = 9$ $s(^-10) = \|^-10\| + 6 = 10 + 6 = 16$ $s(^-6) = \|^-6\| + 6 = 6 + 6 = 12$
$m(x) = x - 1$ $m(8) = 8 - 1 = 7$ $m(^-8) = ^-8 - 1 = ^-9$ $m(0) = 0 - 1 = ^-1$	$n(x) = 1 - x$ $n(8) = 1 - 8 = ^-7$ $n(^-8) = 1 - ^-8 = 9$ $n(0) = 1 - 0 = 1$

Key to Algebra – ANSWERS

Book 5, Page 33

A **table** is an easy way to list the pairs of numbers that belong to a function. The expression for the function is written above the table. The first column of the table lists the numbers to be substituted for the variable in the function. The second column lists the **values** obtained by substituting the numbers in the first column.

$f(x) = x - 10$

x	f(x)
25	25 - 10 = 15
17	17 - 10 = 7
12	12 - 10 = 2
10	10 - 10 = 0
8	8 - 10 = -2
2	2 - 10 = -8
0	0 - 10 = -10
-3	-3 - 10 = -13

$g(x) = 2x + 6$

x	g(x)
-6	2(-6) + 6 = -6
-2	2(-2) + 6 = 2
-1	2(-1) + 6 = 4
0	2(0) + 6 = 6
1	2(1) + 6 = 8
2	2(2) + 6 = 10
6	2(6) + 6 = 18
100	2(100) + 6 = 206

$h(x) = 0 \cdot x$

x	h(x)
-90	$0 \cdot -90 = 0$
-10	$0 \cdot -10 = 0$
-2	$0 \cdot -2 = 0$
-1	$0 \cdot -1 = 0$
0	$0 \cdot 0 = 0$
1	$0 \cdot 1 = 0$
15	$0 \cdot 15 = 0$
40	$0 \cdot 40 = 0$

$k(x) = \frac{x}{6}$

x	k(x)
0	$\frac{0}{6} = 0$
18	$\frac{18}{6} = 3$
-18	$\frac{-18}{6} = -3$
60	$\frac{60}{6} = 10$
-60	$\frac{-60}{6} = -10$
1	$\frac{1}{6}$
-1	$\frac{-1}{6}$
5	$\frac{5}{6}$

$m(x) = \frac{3x}{2}$

x	m(x)
-3	$\frac{3(-3)}{2} = \frac{-9}{2}$
-2	$\frac{3(-2)}{2} = \frac{-6}{2} = -3$
-1	$\frac{3(-1)}{2} = \frac{-3}{2}$
0	$\frac{3(0)}{2} = \frac{0}{2} = 0$
1	$\frac{3(1)}{2} = \frac{3}{2}$
2	$\frac{3(2)}{2} = \frac{6}{2} = 3$
3	$\frac{3(3)}{2} = \frac{9}{2}$
4	$\frac{3(4)}{2} = \frac{12}{2} = 6$

$n(x) = \frac{x-5}{4}$

x	n(x)
-3	$\frac{-3-5}{4} = \frac{-8}{4} = -2$
-2	$\frac{-2-5}{4} = \frac{-7}{4}$
-1	$\frac{-1-5}{4} = \frac{-6}{4}$
0	$\frac{0-5}{4} = \frac{-5}{4}$
1	$\frac{1-5}{4} = \frac{-4}{4} = -1$
2	$\frac{2-5}{4} = \frac{-3}{4}$
3	$\frac{3-5}{4} = \frac{-2}{4}$
4	$\frac{4-5}{2} = \frac{-1}{4}$

Book 5, Page 34

Many functions important in science and other fields have been discovered by people who collected data, organized their information in a table, and figured out a **formula** for it.

See if you can find a formula to fit the data in each table.

$f(x) = 3x$

x	f(x)
5	15
2	6
1	3
0	0
-1	-3
-2	-6
-5	-15

$g(x) = x + 2$

x	g(x)
10	12
5	7
3	5
0	2
-3	-1
-5	-3
-15	-13

$h(x) = x - 4$

x	h(x)
8	4
4	0
2	-2
0	-4
-2	-6
-4	-8
-8	-12

$p(x) = -x$

x	p(x)
-20	20
-12	12
-6	6
0	0
6	-6
12	-12
20	-20

$q(x) = 2x + 1$

x	q(x)
4	9
2	5
1	3
0	1
-1	-1
-2	-3
-4	-7

$r(x) = x^2$

r	r(x)
3	9
2	4
1	1
0	0
-1	1
-2	4
-3	9

$s(x) = -2x$

x	s(x)
5	-10
3	-6
1	-2
-1	2
-3	6
-5	10
-7	14

$v(x) = |x|$

v	v(x)
7	7
4	4
0	0
-2	2
-5	5
-8	8
-60	60

Book 5, Page 35

Answers to Written Work

① $\frac{-7}{1}$ $\frac{5}{2}$ $\frac{0}{1}$ $\frac{6}{10}$ $\frac{-43}{10}$ $\frac{104}{100}$

② $\frac{-15}{4}$ $-3\frac{3}{4}$ -3.75

③ $-9x = 95$; $x = -10\frac{5}{9}$

$\frac{5x}{6} = 14$; $5x = 84$; $x = 16\frac{4}{5}$

$\frac{6x-2}{3} = 7$; $6x - 2 = 21$; $6x = 23$; $x = 3\frac{5}{6}$

$x(x+3) = x^2 - 16$; $x^2 + 3x = x^2 - 16$; $\frac{3x}{3} = \frac{-16}{3}$; $x = -5\frac{1}{3}$

④ $\frac{-4}{10}, \frac{1}{10}$ $\frac{2}{3}, 2\frac{2}{3}$

⑤ Integers greater than 3

Integers between -2 and 4

Rational numbers less than -3

Rational numbers greater than or equal to 4

Rational numbers between -5 and 0

Rational numbers not between -2 and 3

⑥ $x \geq -2$ $x < 2.5$

⑦ $|x+1| = 4$; $x+1 = 4$ or $x+1 = -4$; $x = 3$ or $x = -5$

$3x - 2 \leq 19$; $\frac{3x}{3} \leq \frac{21}{3}$; $x \leq 7$

$-\frac{4x}{3} < 20$; $-4x < 60$; $\frac{-4x}{-4} > \frac{60}{-4}$; $x > -15$

Book 5, Page 35 (cont.)

Answers to Written Work

⑦ continued

$|x-5| = 0$; $x - 5 = 0$; $x = 5$

$6 - 5x > 2$; $-5x > -4$; $\frac{-5x}{-5} < \frac{-4}{-5}$; $x < \frac{4}{5}$

$7x - 1 > 4 - 3x$; $10x - 1 > 4$; $\frac{10x}{10} > \frac{5}{10}$; $x > \frac{1}{2}$

$|x+4| < 12$; $x+4 < 12$ or $x+4 > -12$; $x < 8$ or $x > -16$

$3|x| - 7 = 11$; $\frac{3|x|}{3} = \frac{18}{3}$; $|x| = 6$; $x = 6$ or -6

$3(x-2) \leq 4x + 16$; $3x - 6 \leq 4x + 16$; $-6 \leq x + 16$; $-22 \leq x$; $x \geq -22$

⑧ F , T , F , T , T

⑨ Possible answers are shown below.

f: $f(x) = 2x - 1$

x	f(x)
0	0 - 1 = -1
1	2 - 1 = 1
2	4 - 1 = 3
-1	-2 - 1 = -3
-2	-4 - 1 = -5

g: $g(x) = |x-2|$

x	g(x)
-3	$|-3-2| = |-5| = 5$
-2	$|-2-2| = |-4| = 4$
0	$|0-2| = |-2| = 2$
1	$|1-2| = |-1| = 1$
2	$|2-2| = |0| = 0$

h: $h(x) = x^2$

x	h(x)
1	$1^2 = 1$
2	$2^2 = 4$
3	$3^2 = 9$
4	$4^2 = 16$
5	$5^2 = 25$

Key to Algebra – ANSWERS

Book 5, Page 36

Practice Test

Write each rational number as a fraction.

$-1 = \frac{-1}{1}$ $3\frac{4}{5} = \frac{19}{5}$ $-2\frac{1}{9} = \frac{-19}{9}$ $0.7 = \frac{7}{10}$ $4.01 = \frac{401}{100}$ $0 = \frac{0}{1}$

Draw a number line showing numbers from -3 to 3. Graph the numbers -2.5, 0.3 and $2\frac{1}{3}$.

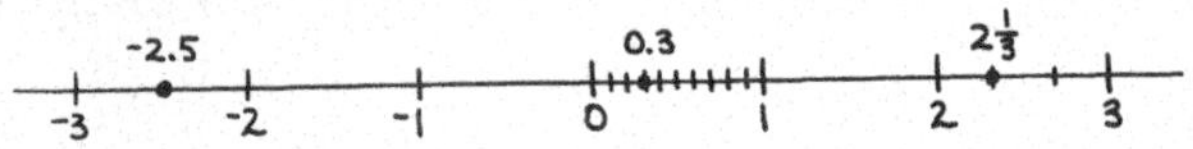

Label each number line and graph the set of numbers.

Integers between -2 and 4:

Rational numbers greater than or equal to 5:

Rational numbers not equal to 1:

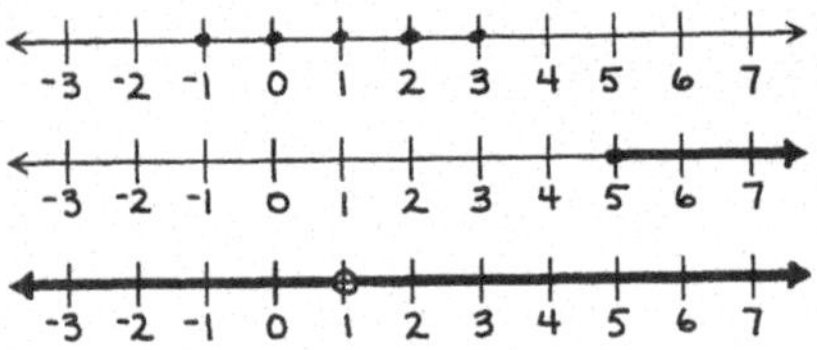

Find each absolute value.

$|8.3| = 8.3$ $|-5| = 5$ $|0| = 0$ $|-2+6| = |4| = 4$

Solve.

$\|x\| = 7$ $x = 7$ or -7	$\|x-5\| = 2$ $x-5 = 2$ or $x-5 = -2$ $x = 7$ or $x = 3$	$\|x\|+4 = 11$ $\|x\| = 7$ $x = 7$ or -7
$2x+3 > 11$ $\frac{2x}{2} > \frac{8}{2}$ $x > 4$	$x-5 < -9$ $x < -4$	$-6x \geq 48$ $\frac{-6x}{-6} \leq \frac{48}{-6}$ $x \leq -8$

Book 5, Page 37

Solve.

$4x-7 < x+8$ $3x-7 < 8$ $3x < 15$ $x < 5$	$9-x \geq -3$ $9 \geq -3+x$ $12 \geq x$ $x \leq 12$	$2 \cdot \frac{x}{2} < -9 \cdot 2$ $x < -18$
$20 < 5x-2$ $\frac{22}{5} < \frac{5x}{5}$ $\frac{22}{5} < x$ $x > \frac{22}{5}$	$7x-3x+2 > x+5x-6$ $4x+2 > 6x-6$ $2 > 2x-6$ $8 > 2x$ $2x < 8$ $x < 4$	$\frac{2x}{5}+6 \leq 4$ $5 \cdot \frac{2x}{5} \leq -2 \cdot 5$ $\frac{2x}{2} \leq \frac{-10}{2}$ $x \leq -5$

$G(x)$ means "a number greater than x." Find a number to make each of these true.

$G(4) = 5$ $G(-2.3) = -2$ $G(0) = 1$

Fill in the missing numbers in the table for each function.

$f(x) = x - 5$

x	$f(x)$
100	$100-5=95$
10	$10-5=5$
4	$4-5=-1$
1	$1-5=-4$
0	$0-5=-5$
-2	$-2-5=-7$
-5	$-5-5=-10$

$g(x) = x^2 + 1$

x	$g(x)$
20	$20^2+1 = 400+1 = 401$
4	$4^2+1 = 16+1 = 17$
1	$1^2+1 = 1+1 = 2$
0	$0^2+1 = 0+1 = 1$
-1	$(-1)^2+1 = 1+1 = 2$
-3	$(-3)^2+1 = 9+1 = 10$
-10	$(-10)^2+1 = 100+1 = 101$

Book 6, Page 1

Review

Remember that a rational number can always be written as a fraction with integers as the numerator and denominator. In this book we will study algebraic expressions that can stand for rational numbers. You will have to use what you know about polynomials as well as what you know about rational numbers.

Write each rational number as a fraction. Then graph it on the number line.

$-4 = \frac{-4}{1}$ $1\frac{2}{5} = \frac{7}{5}$ $3.4 = \frac{34}{10}$ $-2\frac{1}{2} = \frac{-5}{2}$ $-4.9 = \frac{-49}{10}$ $0 = \frac{0}{1}$

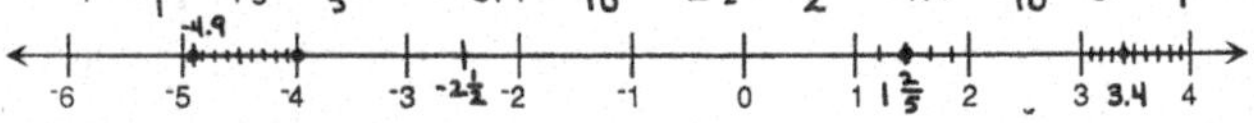

Factor 36 five different ways.

36 → 1 · 36 36 → 2 · 18 36 → 3 · 12 36 → 4 · 9 36 → 6 · 6

Write an equivalent expression using exponents.

$5 \cdot xxxx = 5x^4$ $3n \cdot 5n = 15n^2$

$7y \cdot 7y \cdot 7y = (7y)^3$ or 7^3y^3 $p \cdot 4p \cdot 2p = 8p^3$

$rr \cdot 2t \cdot 2t \cdot 2t = r^2(2t)^3$ or $2^3r^2t^3$ $(xx)(yyy) = x^2y^3$

Multiply.

$c^8 \cdot c = c^9$ $x(x+4) = x^2+4x$

$a^2 \cdot a^5 = a^7$ $5h(h+3) = 5h^2+15h$

$2y^4 \cdot 6y^3 = 12y^7$ $(x+2)(x+6) = x^2+6x+2x+12 = x^2+8x+12$

$6(p-5) = 6p-30$ $(2x+3)(3x-1) = 6x^2-2x+9x-3 = 6x^2+7x-3$

Factor.

$3x-12 = 3(x-4)$ $x^2+6x+9 = (x+3)(x+3)$

$8y-12 = 4(2y-3)$ $t^2-11t+10 = (t-10)(t-1)$

$a^2-81 = (a+9)(a-9)$ $3n^2-n-4 = (3n-4)(n+1)$

Solve each equation.

$5 \cdot \frac{x}{5} = 30 \cdot 5$, $x = 150$ $12 \cdot \frac{5x}{12} = 15 \cdot 12$, $\frac{5x}{5} = \frac{180}{5}$, $x = 36$ $3 \cdot \frac{x-4}{3} = 12 \cdot 3$, $x-4 = 36$, $x = 40$

Book 6, Page 2

Rational Expressions

Remember that a term is an algebraic expression in which multiplication is the only operation. A polynomial can be a single term or can be made by adding and subtracting terms.

Polynomials: x^2+1 19 a^2-4a+5 $3x^3$ $-5x+2y$

A rational number is a fraction with *integers* for the numerator and denominator.
A **rational expression** is a fraction with *polynomials* for the numerator and denominator.

Rational Number: $\frac{\text{Integer}}{\text{Integer}}$ Rational Expression: $\frac{\text{Polynomial}}{\text{Polynomial}}$

Of course, the integer or polynomial on the bottom cannot be 0 (since we cannot divide by 0). Here are some examples of rational expressions:

$\frac{7}{x}$ $\frac{x-5}{x+5}$ $\frac{y+3}{y^2-9}$ $\frac{0}{n+8}$ $\frac{x^2-4x+4}{x+6}$

Make as many rational expressions as you can by using one of these polynomials for the numerator and one for the denominator.

x^2 10 $x-5$ x^2+3x+2

$\frac{x^2}{10}$	$\frac{10}{x^2}$	$\frac{x^2}{x-5}$	$\frac{x-5}{x^2}$
$\frac{x^2}{x^2+3x+2}$	$\frac{x^2+3x+2}{x^2}$	$\frac{10}{x-5}$	$\frac{x-5}{10}$
$\frac{10}{x^2+3x+2}$	$\frac{x^2+3x+2}{10}$	$\frac{x-5}{x^2+3x+2}$	$\frac{x^2+3x+2}{x-5}$
$\frac{x^2}{x^2}$	$\frac{10}{10}$	$\frac{x-5}{x-5}$	$\frac{x^2+3x+2}{x^2+3x+2}$

Key to Algebra – ANSWERS

Book 6, Page 3

Find the value of each rational expression when $x = 3$. If the answer is not an integer, leave it as a fraction.

$\frac{x+4}{x+7} = \frac{3+4}{3+7} = \frac{7}{10}$	$\frac{x}{x+2} = \frac{3}{3+2} = \frac{3}{5}$	$\frac{6}{-x} = \frac{6}{-3} = -2$
$\frac{x-5}{x+3} = \frac{3-5}{3+3} = \frac{-2}{6}$	$\frac{x-1}{x+1} = \frac{3-1}{3+1} = \frac{2}{4}$	$\frac{x^2+1}{x+1} = \frac{3^2+1}{3+1} = \frac{9+1}{4} = \frac{10}{4}$
$\frac{2x-6}{x-6} = \frac{2(3)-6}{3-6} = \frac{0}{-3} = 0$	$\frac{x^2-4}{x+2} = \frac{3^2-4}{3+2} = \frac{9-4}{5} = \frac{5}{5} = 1$	$\frac{10+x}{10-x} = \frac{10+3}{10-3} = \frac{13}{7}$

$\frac{x-8}{x^2+2x+1} = \frac{3-8}{3^2+2(3)+1} = \frac{-5}{9+6+1} = \frac{-5}{16}$	$\frac{x^2-5x+10}{x-5} = \frac{3^2-5(3)+10}{3-5} = \frac{9-15+10}{-2} = \frac{4}{-2} = -2$
$\frac{x^2+4x+4}{x^2-4x+4} = \frac{3^2+4(3)+4}{3^2-4(3)+4} = \frac{9+12+4}{9-12+4} = \frac{25}{1} = 25$	$\frac{-2x^2}{4x-1} = \frac{-2(3)^2}{4(3)-1} = \frac{-2(9)}{12-1} = \frac{-18}{11}$

Find the value of each rational expression when $x = {}^-2$ and $y = 5$.

$\frac{x}{y} = \frac{-2}{5}$	$\frac{3x}{3y} = \frac{3(-2)}{3(5)} = \frac{-6}{15}$	$\frac{y}{x} = \frac{5}{-2}$
$\frac{x+2}{y+2} = \frac{-2+2}{5+2} = \frac{0}{7} = 0$	$\frac{x-2}{y-2} = \frac{-2-2}{5-2} = \frac{-4}{3}$	$\frac{x+5}{y+5} = \frac{-2+5}{5+5} = \frac{3}{10}$
$\frac{1}{xy} = \frac{1}{(-2)(5)} = \frac{1}{-10}$	$\frac{xy}{1} = \frac{(-2)(5)}{1} = \frac{-10}{1} = -10$	$\frac{x^2y^2}{xy} = \frac{(-2)^2(5)^2}{(-2)(5)} = \frac{4 \cdot 25}{-10} = \frac{100}{-10} = -10$

$\frac{x}{2} + \frac{y}{5} = \frac{-2}{2} + \frac{5}{5} = -1 + 1 = 0$	$x\left(\frac{y}{y-7}\right) = -2\left(\frac{5}{5-7}\right) = -2\left(\frac{5}{-2}\right) = 5$

$$\frac{x^2+2xy+y^2}{x+y} = \frac{(-2)^2+2(-2)(5)+5^2}{-2+5} = \frac{4-20+25}{3} = \frac{9}{3} = 3$$

Book 6, Page 4

You might have noticed that every answer on page 3 was a rational number. Once in a while we do *not* get a rational number for an answer when we replace variables with numbers in a rational expression. Look at the substitution table below.

a	$\frac{6}{a+5}$
4	$\frac{6}{4+5} = \frac{6}{9}$
1	$\frac{6}{1+5} = \frac{6}{6} = 1$
0	$\frac{6}{0+5} = \frac{6}{5}$
⁻5	$\frac{6}{-5+5} = \frac{6}{0}$ (crossed out)

Do you see what happened when $a = {}^-5$? When $a = {}^-5$, $\frac{6}{a+5}$ has no value because it equals $6 \div 0$, which has no answer. We say that $\frac{6}{a+5}$ is **undefined** when $a = {}^-5$.

You finish these substitution tables. Be on the lookout for substitutions which give no answer. Write "undefined" whenever you find one.

x	$\frac{x+3}{2}$
3	$\frac{3+3}{2} = \frac{6}{2} = 3$
0	$\frac{0+3}{2} = \frac{3}{2}$
⁻3	$\frac{-3+3}{2} = \frac{0}{2} = 0$
⁻6	$\frac{-6+3}{2} = \frac{-3}{2}$

n	$\frac{n+1}{n+4}$
4	$\frac{4+1}{4+4} = \frac{5}{8}$
2	$\frac{2+1}{2+4} = \frac{3}{6}$
0	$\frac{0+1}{0+4} = \frac{1}{4}$
⁻4	$\frac{-4+1}{-4+4} = \frac{-3}{0}$ undefined

r	$\frac{r}{r^2-9}$
0	$\frac{0}{0^2-9} = \frac{0}{-9} = 0$
1	$\frac{1}{1^2-9} = \frac{1}{1-9} = \frac{1}{-8} = \frac{-1}{8}$
3	$\frac{3}{3^2-9} = \frac{3}{9-9} = \frac{3}{0}$ undefined
⁻3	$\frac{-3}{(-3)^2-9} = \frac{-3}{9-9} = \frac{-3}{0}$ undefined

b	c	$\frac{b}{c}$
5	⁻5	$\frac{5}{-5} = -1$
2	0	$\frac{2}{0}$ undefined
0	2	$\frac{0}{2} = 0$
⁻2	⁻1	$\frac{-2}{-1} = 2$

Book 6, Page 5

To be sure that a rational expression always stands for a number, we have to set the **condition** that the denominator cannot equal zero.

$\frac{x}{y},\ y \neq 0$ (I can choose any number except 0 for y.) $\qquad \frac{a}{a+2},\ a+2 \neq 0 \qquad \frac{x^2+5x+6}{x^2-9},\ x^2-9 \neq 0$

You write the condition for each fraction.

$\frac{a}{b}$, $b \neq 0$ $\qquad \frac{5}{y}$, $y \neq 0$

$\frac{x+y}{x-y}$, $x-y \neq 0$ $\qquad \frac{2x}{3x-5}$, $3x-5 \neq 0$

From now on we shall *assume* that the denominator of each fraction is not zero. That's so we won't have to write down the condition each time.

You already know that every integer can be written as a fraction with a denominator of 1. Every polynomial can also be written as a fraction with a denominator of 1. This means that every polynomial is a rational expression.

$x+3 = \frac{x+3}{1} \qquad a^2-6 = \frac{a^2-6}{1} \qquad y^2+3y-5 = \frac{y^2+3y-5}{1}$

Write each polynomial as a fraction.

$5 - x = \frac{5-x}{1} \qquad 14 - y^2 = \frac{14-y^2}{1}$

$x^2 + 4x = \frac{x^2+4x}{1} \qquad 3a^4 = \frac{3a^4}{1}$

$a^2 - 2 = \frac{a^2-2}{1} \qquad x^3 + x^2 = \frac{x^3+x^2}{1}$

$n^2 + 7n + 12 = \frac{n^2+7n+12}{1} \qquad 5xy = \frac{5xy}{1}$

$5c^2 + 10 = \frac{5c^2+10}{1} \qquad x^2 + 7x + 10 = \frac{x^2+7x+10}{1}$

Book 6, Page 6

Multiplying Fractions

Multiplying fractions is easy, whether they are numbers or expressions. You just multiply the numerators and multiply the denominators.

Multiply.

$\frac{5}{3} \cdot \frac{7}{2} = \frac{35}{6}$	$\frac{4}{11} \cdot \frac{5}{3} = \frac{20}{33}$	$\frac{-6}{7} \cdot \frac{1}{8} = \frac{-6}{56}$	$\frac{9}{10} \cdot \frac{3}{2} = \frac{27}{20}$
$\frac{5}{6} \cdot \frac{-2}{7} = \frac{-10}{42}$	$\frac{-3}{5} \cdot \frac{-3}{5} = \frac{9}{25}$	$\frac{2}{3} \cdot \frac{4}{3} \cdot \frac{2}{5} = \frac{16}{45}$	$\frac{-1}{5} \cdot \frac{3}{2} \cdot \frac{3}{4} = \frac{-9}{40}$
$\frac{x}{3y} \cdot \frac{4}{3y} = \frac{4x}{9y^2}$	$\frac{a}{b} \cdot \frac{a}{b} = \frac{a^2}{b^2}$	$\frac{3}{7} \cdot \frac{x}{y} = \frac{3x}{7y}$	$\frac{4}{a} \cdot \frac{b}{9} = \frac{4b}{9a}$
$\frac{10x}{7} \cdot \frac{2x}{3} = \frac{20x^2}{21}$	$\frac{3x^2}{5y} \cdot \frac{x^3}{2y^2} = \frac{3x^5}{10y^3}$	$\frac{5x^3}{3y^5} \cdot \frac{7x^2}{8y} = \frac{35x^5}{24y^6}$	$\frac{2x}{3y} \cdot \frac{2x}{3y} \cdot \frac{2x}{3y} = \frac{8x^3}{27y^3}$

Multiplying a fraction by an integer or a polynomial is easy, too. Write the integer or the polynomial as a fraction by putting a 1 on the bottom. Then go ahead and multiply numerators and denominators.

$\frac{3}{1} \cdot \frac{2}{5} = \frac{6}{5}$	$\frac{-5}{1} \cdot \frac{3}{4} = \frac{-15}{4}$	$\frac{6}{1} \cdot \frac{-7}{5} = \frac{-42}{5}$	$\frac{1}{5} \cdot \frac{6}{1} = \frac{6}{5}$
$\frac{x}{1} \cdot \frac{x}{y} = \frac{x^2}{y}$	$\frac{5}{1} \cdot \frac{x}{2} = \frac{5x}{2}$	$\frac{x}{1} \cdot \frac{y}{4} = \frac{xy}{4}$	$\frac{3}{5} \cdot \frac{x}{1} = \frac{3x}{5}$
$\frac{3x^2}{1} \cdot \frac{x^2}{y^2} = \frac{3x^4}{y^2}$	$\frac{5a}{3b} \cdot \frac{5a^3}{1} = \frac{25a^4}{3b}$	$\frac{x^3}{1} \cdot \frac{x^2}{y} = \frac{x^5}{y}$	$\frac{2a}{7b} \cdot \frac{3c}{1} = \frac{6ac}{7b}$

Some multiplication problems look hard but are easy if you change each number to a fraction.

$2\frac{5}{8} \cdot 5\frac{1}{3} = \frac{21}{8} \cdot \frac{16}{3} = \frac{14}{1} = 14$	$3\frac{4}{5} \cdot 2\frac{1}{4} = \frac{19}{5} \cdot \frac{9}{4} = \frac{171}{20}$
$5\frac{4}{7} \cdot 1\frac{1}{13} = \frac{39}{7} \cdot \frac{14}{13} = \frac{6}{1} = 6$	$-1\frac{5}{7} \cdot 1\frac{3}{4} = \frac{-12}{7} \cdot \frac{7}{4} = \frac{-3}{1} = -3$
$(3.5a)(\frac{2}{7}b) = \frac{35a}{10} \cdot \frac{2b}{7} = \frac{ab}{1} = ab$	$(2\frac{1}{3})(.09x^2) = \frac{7}{3} \cdot \frac{9x^2}{100} = \frac{21x^2}{100}$

Key to Algebra – ANSWERS

Book 6, Page 7

Multiply.

$\frac{3}{x+5}\cdot\frac{x-2}{x}=\frac{3(x-2)}{x(x+5)}=\frac{3x-6}{x^2+5x}$

$\frac{x}{x-8}\cdot\frac{x+5}{6}=\frac{x(x+5)}{6(x-8)}=\frac{x^2+5x}{6x-48}$

$\frac{7}{x+4}\cdot\frac{x-4}{2x}=\frac{7(x-4)}{2x(x+4)}=\frac{7x-28}{2x^2+8x}$

$\frac{3}{4}\cdot\frac{x+5}{x+7}=\frac{3(x+5)}{4(x+7)}=\frac{3x+15}{4x+28}$

$\frac{x}{y}\cdot\frac{x+y}{x-y}=\frac{x(x+y)}{y(x-y)}=\frac{x^2+xy}{xy-y^2}$

$\frac{2a+3}{3a+4}\cdot\frac{2}{3}=\frac{2(2a+3)}{3(3a+4)}=\frac{4a+6}{9a+12}$

$\frac{5}{x+3}\cdot(x-4)=\frac{5(x-4)}{1(x+3)}=\frac{5x-20}{x+3}$

$(a+5)\cdot\frac{a}{a-2}=\frac{a(a+5)}{1(a-2)}=\frac{a^2+5a}{a-2}$

$\frac{x-3}{2x}\cdot\frac{x-6}{3x}=\frac{(x-3)(x-6)}{2x\cdot 3x}=\frac{x^2-9x+18}{6x^2}$

$\frac{x+3}{x}\cdot\frac{x+4}{x}=\frac{(x+3)(x+4)}{x\cdot x}=\frac{x^2+4x+3x+12}{x^2}=\frac{x^2+7x+12}{x^2}$

$\frac{x+5}{x-6}\cdot\frac{x-6}{x+5}=\frac{(x+5)(x-6)}{(x-6)(x+5)}=\frac{x^2-6x+5x-30}{x^2+5x-6x-30}=\frac{x^2-x-30}{x^2-x-30}=1$

$\frac{a-3}{a-7}\cdot\frac{a+3}{a+7}=\frac{(a-3)(a+3)}{(a-7)(a+7)}=\frac{a^2+3a-3a-9}{a^2+7a-7a-49}=\frac{a^2-9}{a^2-49}$

$\frac{(x+6)}{1}\cdot\frac{x-2}{x-3}=\frac{(x+6)(x-2)}{1(x-3)}=\frac{x^2-2x+6x-12}{x-3}=\frac{x^2+4x-12}{x-3}$

$\frac{y-5}{y+5}\cdot\frac{(y-5)}{1}=\frac{(y-5)(y-5)}{y+5}=\frac{y^2-5y-5y+25}{y+5}=\frac{y^2-10y+25}{y+5}$

Book 6, Page 8

Here are some rational expressions which have exponents. Write each one out the long way. Then multiply.

$\left(\frac{2}{5}\right)^2=\frac{2}{5}\cdot\frac{2}{5}=\frac{4}{25}$

$\left(\frac{3}{4}\right)^2=\frac{3}{4}\cdot\frac{3}{4}=\frac{9}{16}$

$\left(\frac{-1}{6}\right)^2=\frac{-1}{6}\cdot\frac{-1}{6}=\frac{1}{36}$

$\left(\frac{x}{y}\right)^2=\frac{x}{y}\cdot\frac{x}{y}=\frac{x^2}{y^2}$

$\left(\frac{2x}{9}\right)^2=\frac{2x}{9}\cdot\frac{2x}{9}=\frac{4x^2}{81}$

$\left(\frac{3a}{10}\right)^2=\frac{3a}{10}\cdot\frac{3a}{10}=\frac{9a^2}{100}$

$\left(\frac{-2}{3}\right)^3=\frac{-2}{3}\cdot\frac{-2}{3}\cdot\frac{-2}{3}=\frac{-8}{27}$

$\left(\frac{-2}{3}\right)^4=\frac{-2}{3}\cdot\frac{-2}{3}\cdot\frac{-2}{3}\cdot\frac{-2}{3}=\frac{16}{81}$

$\left(\frac{x}{y}\right)^4=\frac{x}{y}\cdot\frac{x}{y}\cdot\frac{x}{y}\cdot\frac{x}{y}=\frac{x^4}{y^4}$

$\left(\frac{3a}{10}\right)^3=\frac{3a}{10}\cdot\frac{3a}{10}\cdot\frac{3a}{10}=\frac{27a^3}{1000}$

$\left(\frac{x-5}{x+6}\right)^2=\frac{x-5}{x+6}\cdot\frac{x-5}{x+6}=\frac{x^2-10x+25}{x^2+12x+36}$

$\left(\frac{x+2}{x+4}\right)^2=\frac{x+2}{x+4}\cdot\frac{x+2}{x+4}=\frac{x^2+4x+4}{x^2+8x+16}$

$\left(\frac{x-3}{x-7}\right)^2=\frac{x-3}{x-7}\cdot\frac{x-3}{x-7}=\frac{x^2-6x+9}{x^2-14x+49}$

$\left(\frac{5}{x+5}\right)^2=\frac{5}{x+5}\cdot\frac{5}{x+5}=\frac{25}{x^2+10x+25}$

$\left(\frac{2x+1}{x-9}\right)^2=\frac{2x+1}{x-9}\cdot\frac{2x+1}{x-9}=\frac{4x^2+4x+1}{x^2-18x+81}$

Book 6, Page 9

Equivalent Fractions

Every fraction that has the same numerator and denominator (except $\frac{0}{0}$) is equal to 1. That's because the denominator goes into the numerator one time. Each fraction below is equal to 1.

$\frac{2}{2}\quad\frac{3}{3}\quad\frac{4}{4}\quad\frac{8}{8}\quad\frac{100}{100}\quad\frac{-1}{-1}\quad\frac{-2}{-2}\quad\frac{-5}{-5}\quad\frac{-50}{-50}$

When we multiply a number by 1, we always end up with a number equal to the number we started with. For example:

$5\cdot 1=5\qquad -6\cdot 1=-6\qquad 342\cdot 1=342$

The same thing happens when we multiply a fraction by another fraction that's equal to 1. We end up with a fraction that's **equivalent** to the fraction we started with. Here are some examples:

$\frac{1}{2}\cdot\frac{2}{2}=\frac{2}{4}\qquad\frac{1}{2}\cdot\frac{3}{3}=\frac{3}{6}\qquad\frac{1}{2}\cdot\frac{4}{4}=\frac{4}{8}$

(equivalent fractions) (equivalent fractions) (equivalent fractions)

$\frac{2}{4}$, $\frac{3}{6}$ and $\frac{4}{8}$ are all equivalent to $\frac{1}{2}$. Find some more fractions equivalent to $\frac{1}{2}$.

$\frac{1}{2}\cdot\frac{5}{5}=\frac{5}{10}\qquad\frac{1}{2}\cdot\frac{6}{6}=\frac{6}{12}\qquad\frac{1}{2}\cdot\frac{7}{7}=\frac{7}{14}\qquad\frac{1}{2}\cdot\frac{8}{8}=\frac{8}{16}$

$\frac{1}{2}\cdot\frac{9}{9}=\frac{9}{18}\qquad\frac{1}{2}\cdot\frac{-9}{-9}=\frac{-9}{-18}\qquad\frac{1}{2}\cdot\frac{50}{50}=\frac{50}{100}\qquad\frac{1}{2}\cdot\frac{100}{100}=\frac{100}{200}$

(You pick the number.)

Find some fractions equivalent to $\frac{3}{4}$.

$\frac{3}{4}\cdot\frac{2}{2}=\frac{6}{8}\qquad\frac{3}{4}\cdot\frac{3}{3}=\frac{9}{12}\qquad\frac{3}{4}\cdot\frac{4}{4}=\frac{12}{16}\qquad\frac{3}{4}\cdot\frac{-5}{-5}=\frac{-15}{-20}$

$\frac{3}{4}\cdot\frac{-8}{-8}=\frac{-24}{-32}\qquad\frac{3}{4}\cdot\frac{10}{10}=\frac{30}{40}\qquad\frac{3}{4}\cdot\frac{20}{20}=\frac{60}{80}\qquad\frac{3}{4}\cdot\frac{25}{25}=\frac{75}{80}$

(You pick numbers for these.)

Book 6, Page 10

Label the points shown on each number line below.

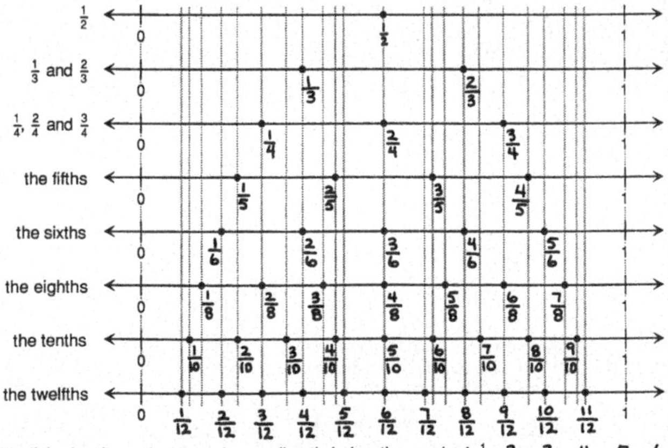

List all the fractions whose graphs are directly below the graph of $\frac{1}{2}$: $\frac{2}{4},\frac{3}{6},\frac{4}{8},\frac{5}{10},\frac{6}{12}$

What is true about all of these fractions?

They are all equivalent to $\frac{1}{2}$.

List all the fractions shown above that are equivalent to each fraction below.

$\frac{1}{3}$: $\frac{2}{6},\frac{4}{12}$ $\qquad\frac{3}{4}$: $\frac{6}{8},\frac{9}{12}$ $\qquad\frac{4}{5}$: $\frac{8}{10}$

$\frac{2}{3}$: $\frac{4}{6},\frac{8}{12}$ $\qquad\frac{1}{5}$: $\frac{2}{10}$ $\qquad\frac{1}{6}$: $\frac{2}{12}$

$\frac{1}{4}$: $\frac{2}{8},\frac{3}{12}$ $\qquad\frac{2}{5}$: $\frac{4}{10}$ $\qquad\frac{2}{6}$: $\frac{1}{3},\frac{4}{12}$

$\frac{2}{4}$: $\frac{1}{2},\frac{3}{6},\frac{4}{8},\frac{5}{10},\frac{6}{12}$ $\qquad\frac{3}{5}$: $\frac{6}{10}$ $\qquad\frac{5}{6}$: $\frac{10}{12}$

Key to Algebra – ANSWERS

Book 6, Page 11

Rewriting Fractions in Higher Terms

As you can see, there is an easy way to rewrite a fraction in higher terms. We just pick a number or expression which is not 0 and multiply the numerator and denominator of the fraction by the number or expression we picked. That amounts to multiplying the fraction by 1, so the answer comes out equivalent to the fraction we started with.

Pick 3: $\frac{7}{8} = \frac{7 \cdot 3}{8 \cdot 3} = \frac{21}{24}$

We ended up with $\frac{21}{24}$, which is equivalent to $\frac{7}{8}$ and is in **higher terms**.

For each fraction, find an equivalent fraction in higher terms.

Multiply numerator and denominator			
by 5:	$\frac{5 \cdot 5}{6 \cdot 5} = \frac{25}{30}$	$\frac{-2 \cdot 5}{7 \cdot 5} = \frac{-10}{35}$	$\frac{x \cdot 5}{3 \cdot 5} = \frac{5x}{15}$
by 2:	$\frac{9 \cdot 2}{10 \cdot 2} = \frac{18}{20}$	$\frac{-1 \cdot 2}{3 \cdot 2} = \frac{-2}{6}$	$\frac{4 \cdot 2}{y \cdot 2} = \frac{8}{2y}$
by -3:	$\frac{7 \cdot -3}{2 \cdot -3} = \frac{-21}{-6}$	$\frac{-4 \cdot -3}{5 \cdot -3} = \frac{12}{-15}$	$\frac{(x-1) \cdot -3}{6 \cdot -3} = \frac{-3x+3}{-18}$
by x:	$\frac{3 \cdot x}{8 \cdot x} = \frac{3x}{8x}$	$\frac{1 \cdot x}{6 \cdot x} = \frac{x}{6x}$	$\frac{2x \cdot x}{3y \cdot x} = \frac{2x^2}{3xy}$
by a:	$\frac{9 \cdot a}{5 \cdot a} = \frac{9a}{5a}$	$\frac{-1 \cdot a}{a \cdot a} = \frac{-a}{a^2}$	$\frac{a^2 \cdot a}{(a+4) \cdot a} = \frac{a^3}{a^2+4a}$
by $2n$:	$\frac{1 \cdot 2n}{8 \cdot 2n} = \frac{2n}{16n}$	$\frac{3n^2 \cdot 2n}{m \cdot 2n} = \frac{6n^3}{2mn}$	$\frac{n+1 \cdot 2n}{n^2 \cdot 2n} = \frac{2n^2+2n}{2n^3}$
by $-4b$:	$\frac{2 \cdot -4b}{3 \cdot -4b} = \frac{-8b}{-12b}$	$\frac{5 \cdot -4b}{b \cdot -4b} = \frac{-20b}{-4b^2}$	$\frac{(1-b) \cdot -4b}{b \cdot -4b} = \frac{-4b+4b^2}{-4b^2}$

> To rewrite a fraction in higher terms, multiply the numerator and denominator by the same number or polynomial.

Book 6, Page 12

Find five fractions equivalent to $\frac{2}{5}$.	Find five fractions equivalent to $\frac{1}{4}$.	Find five fractions equivalent to $\frac{3}{8}$.
$\frac{2 \cdot 2}{5 \cdot 2} = \frac{4}{10}$	$\frac{1 \cdot 2}{4 \cdot 2} = \frac{2}{8}$	$\frac{3 \cdot 2}{8 \cdot 2} = \frac{6}{16}$
$\frac{2 \cdot 3}{5 \cdot 3} = \frac{6}{15}$	$\frac{1 \cdot 3}{4 \cdot 3} = \frac{3}{12}$	$\frac{3 \cdot 5}{8 \cdot 5} = \frac{15}{40}$
$\frac{2 \cdot 5}{5 \cdot 5} = \frac{10}{25}$	$\frac{1 \cdot 10}{4 \cdot 10} = \frac{10}{40}$	$\frac{3 \cdot -3}{8 \cdot -3} = \frac{-9}{-24}$
$\frac{2 \cdot 10}{5 \cdot 10} = \frac{20}{50}$	$\frac{1 \cdot -10}{4 \cdot -10} = \frac{-10}{-40}$	$\frac{3 \cdot -8}{8 \cdot -8} = \frac{-24}{-64}$

Find five fractions equivalent to $\frac{2x}{3}$.	Find five fractions equivalent to $\frac{-5}{x}$.	Find five fractions equivalent to $\frac{x}{y}$.
$\frac{2x \cdot 2}{3 \cdot 2} = \frac{4x}{6}$	$\frac{-5 \cdot 2}{x \cdot 2} = \frac{-10}{2x}$	$\frac{x \cdot 2}{y \cdot 2} = \frac{2x}{2y}$
$\frac{2x \cdot 5}{3 \cdot 5} = \frac{10x}{15}$	$\frac{-5 \cdot 10}{x \cdot 10} = \frac{-50}{10x}$	$\frac{x \cdot -6}{y \cdot -6} = \frac{-6x}{-6y}$
$\frac{2x \cdot x}{3 \cdot x} = \frac{2x^2}{3x}$	$\frac{-5 \cdot x}{x \cdot x} = \frac{-5x}{x^2}$	$\frac{x \cdot x}{y \cdot x} = \frac{x^2}{xy}$
$\frac{2x \cdot 3x}{3 \cdot 3x} = \frac{6x^2}{9x}$	$\frac{-5 \cdot (x+4)}{x \cdot (x+4)} = \frac{-5x-20}{x^2+4x}$	$\frac{x \cdot (x+1)}{y \cdot (x+1)} = \frac{x^2+x}{xy+y}$
$\frac{2x \cdot y}{3 \cdot y} = \frac{2xy}{3y}$	$\frac{-5 \cdot (x-1)}{x \cdot (x-1)} = \frac{-5x+5}{x^2-x}$	$\frac{x \cdot (x+2)}{y \cdot (x+2)} = \frac{x^2+2x}{xy+2y}$

Find 5 fractions equivalent to $\frac{x+3}{x-2}$.	Find 5 fractions equivalent to $\frac{x-6}{x+4}$.
$\frac{(x+3) \cdot 2}{(x-2) \cdot 2} = \frac{2x+6}{2x-4}$	$\frac{(x-6) \cdot 2}{(x+4) \cdot 2} = \frac{2x-12}{2x+8}$
$\frac{(x+3) \cdot 5}{(x-2) \cdot 5} = \frac{5x+15}{5x-10}$	$\frac{(x-6) \cdot x}{(x+4) \cdot x} = \frac{x^2-6x}{x^2+4x}$
$\frac{(x+3) \cdot x}{(x-2) \cdot x} = \frac{x^2+3x}{x^2-2x}$	$\frac{(x-6)(x-1)}{(x+4)(x-1)} = \frac{x^2-1x-6x+6}{x^2-1x+4x-4} = \frac{x^2-7x+6}{x^2+3x-4}$
$\frac{(x+3) \cdot 3x}{(x-2) \cdot 3x} = \frac{3x^2+9x}{3x^2-6x}$	$\frac{(x-6)(x+3)}{(x+4)(x+3)} = \frac{x^2+3x-6x-18}{x^2+3x+4x+12} = \frac{x^2-3x-18}{x^2+7x+12}$
$\frac{(x+3)(x+4)}{(x-2)(x+4)} = \frac{x^2+4x+3x+12}{x^2+4x-2x-8} = \frac{x^2+7x+12}{x^2+2x-8}$	$\frac{(x-6)(x+6)}{(x+4)(x+6)} = \frac{x^2+6x-6x-36}{x^2+6x+4x+24} = \frac{x^2-36}{x^2+10x+24}$

Book 6, Page 13

If we want an equivalent fraction with a certain denominator, we must figure out what to multiply the top and bottom by to get that denominator Here are two examples:

$\frac{2}{3} = \frac{}{24}$ (24 is 3 times 8, so I should also multiply 2 by 8.) $\frac{2 \cdot 8}{3 \cdot 8} = \frac{16}{24}$

$\frac{6}{x} = \frac{}{3x^2}$ ($3x^2$ is $x \cdot 3x$, so I should also multiply 6 by $3x$.) $\frac{6 \cdot 3x}{x \cdot 3x} = \frac{18x}{3x^2}$

You try these. First figure out what the denominator has been multiplied by. Then multiply the numerator by the same number or expression.

$\frac{4 \cdot 10}{5 \cdot 10} = \frac{40}{50}$	$\frac{-3 \cdot 7}{5 \cdot 7} = \frac{-21}{35}$	$\frac{7 \cdot 25}{2 \cdot 25} = \frac{175}{50}$
$\frac{-1 \cdot 3}{9 \cdot 3} = \frac{-3}{27}$	$\frac{5 \cdot -4}{-8 \cdot -4} = \frac{-20}{32}$	$\frac{-4 \cdot 21}{3 \cdot 21} = \frac{84}{63}$
$\frac{x \cdot 6}{y \cdot 6} = \frac{6x}{6y}$	$\frac{2t \cdot 4}{3 \cdot 4} = \frac{8t}{12}$	$\frac{n \cdot 5}{2m \cdot 5} = \frac{5n}{10m}$
$\frac{2a \cdot b}{b \cdot b} = \frac{2ab}{b^2}$	$\frac{5p \cdot p^2}{4 \cdot p^2} = \frac{5p^3}{4p^2}$	$\frac{r \cdot -4r}{-2r \cdot -4r} = \frac{-4r^2}{8r^2}$
$\frac{(x+5) \cdot 7}{4 \cdot 7} = \frac{7x+35}{28}$	$\frac{3 \cdot 5}{(x-1) \cdot 5} = \frac{15}{5(x-1)}$	$\frac{(x+1) \cdot 6}{(x+2) \cdot 6} = \frac{6(x+1)}{6(x+2)}$

In these problems you will have to factor the new denominator to find out what to multiply by.

(3x-6 equals 3(x-2), so I should multiply the top and bottom by 3.)

$\frac{7 \cdot 3}{(x-2) \cdot 3} = \frac{21}{\cancel{3x-6}\ 3(x-2)}$	$\frac{6 \cdot 2}{(x-9) \cdot 2} = \frac{12}{\cancel{2x-18}\ 2(x-9)}$
$\frac{5 \cdot 10}{(x+4) \cdot 10} = \frac{50}{\cancel{10x+40}\ 10(x+4)}$	$\frac{1 \cdot 4}{(3x+5) \cdot 4} = \frac{4}{\cancel{12x+20}\ 4(3x+5)}$
$\frac{x \cdot x}{(x-3) \cdot x} = \frac{x^2}{\cancel{x^2-3x}\ x(x-3)}$	$\frac{2n \cdot n}{(n+1) \cdot n} = \frac{2n^2}{\cancel{n^2+n}\ n(n+1)}$
$\frac{2 \cdot (x+2)}{(x+5) \cdot (x+2)} = \frac{2(x+2)}{\cancel{x^2+7x+10}\ (x+5)(x+2)}$	$\frac{3 \cdot (y+4)}{(y-4) \cdot (y+4)} = \frac{3y+12}{\cancel{y^2-16}\ (y+4)(y-4)}$

Book 6, Page 14

Simplifying Fractions

We know how to rewrite a fraction in higher terms. We just multiply the numerator and denominator by the same number.

$$\frac{4 \cdot 5}{7 \cdot 5} = \frac{20}{35}$$

rewritten in higher terms

To **simplify** a fraction, or rewrite it in **lower terms**, we do the opposite. We factor the numerator and denominator of the fraction so that one of the factors on top is the same as one of the factors on the bottom. Then we cancel the same factor from the top and bottom. What is left is our simplified fraction, which will be equivalent to the fraction we started with. Here is an example:

(5 goes into both 20 and 35.)

$$\frac{20}{35} = \frac{5 \cdot 4}{5 \cdot 7} = \frac{\cancel{5}}{\cancel{5}} \cdot \frac{4}{7} = \frac{4}{7}$$

rewritten in lower terms

Simplify each fraction. Do each problem in two steps.

(4 goes into 8 and 12.) (3 is a factor of -3 and 12.)

$\frac{8}{12} = \frac{\cancel{4} \cdot 2}{\cancel{4} \cdot 3} = \frac{2}{3}$	$\frac{-3}{12} = \frac{\cancel{3} \cdot -1}{\cancel{3} \cdot 4} = \frac{-1}{4}$
$\frac{9}{12} = \frac{\cancel{3} \cdot 3}{\cancel{3} \cdot 4} = \frac{3}{4}$	$\frac{-15}{25} = \frac{\cancel{5} \cdot -3}{\cancel{5} \cdot 5} = \frac{-3}{5}$
$\frac{6}{18} = \frac{\cancel{6} \cdot 1}{\cancel{6} \cdot 3} = \frac{1}{3}$	$\frac{-4}{14} = \frac{\cancel{2} \cdot -2}{\cancel{2} \cdot 7} = \frac{-2}{7}$
$\frac{-12}{21} = \frac{\cancel{3} \cdot -4}{\cancel{3} \cdot 7} = \frac{-4}{7}$	$\frac{15}{6} = \frac{\cancel{3} \cdot 5}{\cancel{3} \cdot 2} = \frac{5}{2}$
$\frac{20}{22} = \frac{\cancel{2} \cdot 10}{\cancel{2} \cdot 11} = \frac{10}{11}$	$\frac{18}{24} = \frac{\cancel{6} \cdot 3}{\cancel{6} \cdot 4} = \frac{3}{4}$
$\frac{-4}{16} = \frac{\cancel{4} \cdot -1}{\cancel{4} \cdot 4} = \frac{-1}{4}$	$\frac{-80}{30} = \frac{\cancel{10} \cdot -8}{\cancel{10} \cdot 3} = \frac{-8}{3}$

Key to Algebra – ANSWERS

Book 6, Page 15

Simplify each rational expression by factoring the top and bottom and canceling the common factors.

2 is the only factor left on the top. On the bottom I have 3, y and y. 3·y·y = 3y².

To keep track of the factors that are left, I circled them.

$\frac{6y^3}{9y^5} = \frac{2\cdot\cancel{3}\cdot\cancel{y}\cdot\cancel{y}\cdot\cancel{y}}{3\cdot\cancel{3}\cdot\cancel{y}\cdot\cancel{y}\cdot\cancel{y}\cdot y\cdot y} = \frac{2}{3y^2}$

$\frac{10x^3y^2}{35xy^3} = \frac{(2)\cdot\cancel{5}\cdot\cancel{x}\cdot(x)\cdot(x)\cdot\cancel{y}\cdot\cancel{y}}{(7)\cdot\cancel{5}\cdot\cancel{x}\cdot(y)\cdot\cancel{y}\cdot\cancel{y}} = \frac{2x^2}{7y}$

$\frac{20xy}{5x^2} = \frac{2\cdot2\cdot\cancel{5}\cdot\cancel{x}\cdot y}{\cancel{5}\cdot\cancel{x}\cdot x} = \frac{4y}{x}$

$\frac{12a^4}{10a^3} = \frac{\cancel{2}\cdot(2)\cdot(3)\cdot\cancel{a}\cdot\cancel{a}\cdot\cancel{a}\cdot(a)}{\cancel{2}\cdot(5)\cdot\cancel{a}\cdot\cancel{a}\cdot\cancel{a}} = \frac{6a}{5}$

$\frac{15x^5}{25x^2} = \frac{3\cdot\cancel{5}\cdot\cancel{x}\cdot\cancel{x}\cdot x\cdot x\cdot x}{\cancel{5}\cdot5\cdot\cancel{x}\cdot\cancel{x}} = \frac{3x^3}{5}$

$\frac{-9x^2}{6x^2} = \frac{-1\cdot3\cdot\cancel{3}\cdot\cancel{x}\cdot\cancel{x}}{2\cdot\cancel{3}\cdot\cancel{x}\cdot\cancel{x}} = \frac{-3}{2}$

$\frac{3a^2}{a^5} = \frac{3\cdot\cancel{a}\cdot\cancel{a}}{a\cdot a\cdot a\cdot\cancel{a}\cdot\cancel{a}} = \frac{3}{a^3}$

$\frac{x^2}{xy} = \frac{\cancel{x}\cdot x}{\cancel{x}\cdot y} = \frac{x}{y}$

$\frac{10a^2}{40a} = \frac{\cancel{2}\cdot\cancel{5}\cdot\cancel{a}\cdot a}{\cancel{2}\cdot2\cdot2\cdot\cancel{5}\cdot\cancel{a}} = \frac{a}{4}$

$\frac{5ab}{a^3b} = \frac{5\cdot\cancel{a}\cdot\cancel{b}}{\cancel{a}\cdot a\cdot a\cdot\cancel{b}} = \frac{5}{a^2}$

$\frac{15x^5}{13x^2} = \frac{3\cdot5\cdot\cancel{x}\cdot\cancel{x}\cdot x\cdot x\cdot x}{13\cdot\cancel{x}\cdot\cancel{x}} = \frac{15x^3}{13}$

$\frac{x^2y^2}{x^2yz} = \frac{\cancel{x}\cdot\cancel{x}\cdot\cancel{y}\cdot y}{\cancel{x}\cdot\cancel{x}\cdot\cancel{y}\cdot z} = \frac{y}{z}$

$\frac{9x^2y}{10y^2} = \frac{3\cdot3\cdot x\cdot x\cdot\cancel{y}}{2\cdot5\cdot y\cdot\cancel{y}} = \frac{9x^2}{10y}$

$\frac{x^4y}{xy^4} = \frac{\cancel{x}\cdot x\cdot x\cdot x\cdot\cancel{y}}{\cancel{x}\cdot y\cdot y\cdot y\cdot\cancel{y}} = \frac{x^3}{y^3}$

$\frac{3x^2y^2}{21xy} = \frac{\cancel{3}\cdot\cancel{x}\cdot(x)\cdot(y)\cdot\cancel{y}}{\cancel{3}\cdot(7)\cdot\cancel{x}\cdot\cancel{y}} = \frac{xy}{7}$

$\frac{20x^2y}{30xy^2} = \frac{\cancel{2}\cdot(2)\cdot\cancel{5}\cdot\cancel{x}\cdot(x)\cdot\cancel{y}}{\cancel{2}\cdot(3)\cdot\cancel{5}\cdot\cancel{x}\cdot(y)\cdot\cancel{y}} = \frac{2x}{3y}$

$\frac{27x^3y}{36xy^3} = \frac{\cancel{3}\cdot(3)\cdot\cancel{3}\cdot\cancel{x}\cdot(x)\cdot(x)\cdot\cancel{y}}{(2)\cdot\cancel{3}\cdot(2)\cdot\cancel{3}\cdot\cancel{x}\cdot(y)\cdot(y)\cdot\cancel{y}} = \frac{3x^2}{4y^2}$

$\frac{-60a^2b^2}{40ab} = \frac{(-1)\cdot\cancel{2}\cdot\cancel{2}\cdot(3)\cdot\cancel{5}\cdot\cancel{a}\cdot(a)\cdot\cancel{b}\cdot(b)}{\cancel{2}\cdot\cancel{2}\cdot\cancel{5}\cdot(2)\cdot\cancel{a}\cdot\cancel{b}} = \frac{-3ab}{2}$

Book 6, Page 16

The more factoring and canceling you can do in your head, the less you will have to write out each time you simplify a fraction.

5 goes into 10 and 35. 5 into 10 is 2 and 5 into 35 is 7.

The x on the bottom cancels one of the x's on the top. So x² is left on the top.

The y² on top cancels two of the y's on the bottom. So y is left on the bottom.

$\frac{10x^3y^2}{35xy^3} = \frac{2x^2}{7y}$

Here are some more fractions for you to simplify. Try to do the canceling without writing out all the factors.

$\frac{6y^3}{9y^5} = \frac{2}{3y^2}$	$\frac{20xy}{5x^2} = \frac{4y}{x}$	$\frac{12a^4}{10a^3} = \frac{6a}{5}$
$\frac{15x^5}{25x^2} = \frac{3x^3}{5}$	$\frac{-9x^2}{6x^2} = \frac{-3}{2}$	$\frac{3a^2}{a^5} = \frac{3}{a^3}$
$\frac{x^2}{xy} = \frac{x}{y}$	$\frac{9x^2y}{10y^2} = \frac{9x^2}{10y}$	$\frac{x^4y}{xy^4} = \frac{x^3}{y^3}$
$\frac{3x^2y^2}{21xy} = \frac{xy}{7}$	$\frac{20x^2y}{30xy^2} = \frac{2x}{3y}$	$\frac{27x^3y}{36xy^3} = \frac{3x^2}{4y^2}$
$\frac{27x^3y^3}{36x^3y^3} = \frac{3}{4}$	$\frac{-60a^2b^2}{40ab} = \frac{-3ab}{2}$	$\frac{10a^2}{40a^2} = \frac{a}{4}$
$\frac{ab}{a^3b} = \frac{1}{a^2}$	$\frac{15x^5}{3x^2} = \frac{5x^3}{1} = 5x^3$	$\frac{x^2y}{x^2y} = \frac{1}{1} = 1$

Did you get the last three problems? If you're not sure, look at the next page.

Book 6, Page 17

Sometimes you can cancel out all the factors on the TOP . . .

ab is really 1ab.

$\frac{ab}{a^3b} = \frac{1}{a^2}$	$\frac{a}{a^2} = \frac{1}{a}$	$\frac{x}{5x} = \frac{1}{5}$
$\frac{3x}{9x} = \frac{1}{3}$	$\frac{xy}{x^2y^2} = \frac{1}{xy}$	$\frac{2x^2}{10x^2y} = \frac{1}{5y}$

Sometimes you can cancel out all the factors on the BOTTOM . . .

$\frac{15x^5}{3x^2} = \frac{5x^3}{1} = 5x^3$	$\frac{6x^3}{3x} = \frac{2x^2}{1} = 2x^2$	$\frac{5ab}{a} = \frac{5b}{1} = 5b$
$\frac{xy}{x} = \frac{y}{1} = y$	$\frac{12x^3}{4x^2} = \frac{3x}{1} = 3x$	$\frac{3x^2y^2}{x^2y} = \frac{3y}{1} = 3y$

And sometimes you can cancel out all the factors on the TOP AND BOTTOM . . .

$\frac{x^2y}{x^2y} = \frac{1}{1} = 1$	$\frac{a}{a} = \frac{1}{1} = 1$	$\frac{-3x}{-3x} = \frac{1}{1} = 1$
$\frac{a^2b^2}{a^2b^2} = \frac{1}{1} = 1$	$\frac{4x^2}{4x^2} = \frac{1}{1} = 1$	$\frac{12xy}{12xy} = \frac{1}{1} = 1$

Book 6, Page 18

Simplify these rational expressions. Each one is already factored for you. You just have to cancel and write the answer.

$\frac{5x^2}{10x+15} = \frac{\cancel{5}\cdot x^2}{\cancel{5}(2x+3)} = \frac{x^2}{2x+3}$	$\frac{2a^2}{2a+10} = \frac{\cancel{2}\cdot a^2}{\cancel{2}(a+5)} = \frac{a^2}{a+5}$
$\frac{4x+12}{8x-4y} = \frac{\cancel{4}(x+3)}{\cancel{4}(2x-y)} = \frac{x+3}{2x-y}$	$\frac{6x^2+9x}{3xy-12x^2} = \frac{\cancel{3x}(2x+3)}{\cancel{3x}(y-4x)} = \frac{2x+3}{y-4x}$
$\frac{x^2+3x}{5x+15} = \frac{x\cancel{(x+3)}}{5\cancel{(x+3)}} = \frac{x}{5}$	$\frac{x+4}{2x+8} = \frac{1\cancel{(x+4)}}{2\cancel{(x+4)}} = \frac{1}{2}$
$\frac{3x-12}{x^2-4x} = \frac{3\cancel{(x-4)}}{x\cancel{(x-4)}} = \frac{3}{x}$	$\frac{xy+2y}{7x+14} = \frac{y\cancel{(x+2)}}{7\cancel{(x+2)}} = \frac{y}{7}$

You will have to factor the polynomials in each rational expression below. Then cancel factors and write your answer. Show each step.

$\frac{3y}{3y+6} = \frac{\cancel{3}\cdot y}{\cancel{3}(y+2)} = \frac{y}{y+2}$	$\frac{x-6}{2x^2-12x} = \frac{1\cancel{(x-6)}}{2x\cancel{(x-6)}} = \frac{1}{2x}$
$\frac{x^2+2x}{5x+10} = \frac{x\cancel{(x+2)}}{5\cancel{(x+2)}} = \frac{x}{5}$	$\frac{2x-8}{3x-12} = \frac{2\cancel{(x-4)}}{3\cancel{(x-4)}} = \frac{2}{3}$
$\frac{x^2+5x}{6x+30} = \frac{x\cancel{(x+5)}}{6\cancel{(x+5)}} = \frac{x}{6}$	$\frac{10x}{5x+40} = \frac{\cancel{5}\cdot2x}{\cancel{5}(x+8)} = \frac{2x}{x+8}$
$\frac{x^2-9x}{x^2+xy} = \frac{\cancel{x}(x-9)}{\cancel{x}(x+y)} = \frac{x-9}{x+y}$	$\frac{x^2+4x}{4x+16} = \frac{x\cancel{(x+4)}}{4\cancel{(x+4)}} = \frac{x}{4} = \frac{x}{4}$

Key to Algebra – ANSWERS

Book 6, Page 19

When we are simplifying rational expressions, we can only cancel an expression that is a *factor* of both the top and bottom. An expression is a factor of the top if it is multiplied by all the rest of the top. It is a factor of the bottom if it is multiplied by all the rest of the bottom. Here is how Sandy and Terry did the last problem on page 16.

Sandy	*Terry*
$\frac{x^2+4x}{4x+16} = \frac{x\cancel{(x+4)}}{4\cancel{(x+4)}} = \frac{x}{4}$ C	$\frac{x^2+\cancel{4x}}{\cancel{4x}+16} = \frac{x^2}{16}$ X
Right! Sandy remembered to factor first. $x+4$ is a factor of both top and bottom.	Wrong! $4x$ is not a factor of either the top or the bottom.

Simplify each fraction. Remember to *factor* before you cancel.

$\frac{a^2+3a}{3a+9} = \frac{a(a+3)}{3(a+3)} = \frac{a}{3}$	$\frac{3a^2}{2a+5ab} = \frac{3\cdot a\cdot\cancel{a}}{\cancel{a}(2+5b)} = \frac{3a}{2+5b}$
$\frac{3x+5y}{6x+10y} = \frac{\overset{1}{\cancel{3x+5y}}}{2\cancel{(3x+5y)}} = \frac{1}{2}$	$\frac{x^2+xy}{9x} = \frac{\cancel{x}(x+y)}{9\cancel{x}} = \frac{x+y}{9}$
$\frac{4x+8}{4x+12} = \frac{\cancel{4}(x+2)}{\cancel{4}(x+3)} = \frac{x+2}{x+3}$	$\frac{2x+4}{2x+4} = \frac{1\cancel{(2x+4)}}{1\cancel{(2x+4)}} = \frac{1}{1} = 1$
$\frac{3a}{6a^2+9a} = \frac{\overset{1}{\cancel{3a}}}{\cancel{3a}(2a+3)} = \frac{1}{2a+3}$	$\frac{2a^2+2ab}{4ab} = \frac{\cancel{2a}(a+b)}{\cancel{2}\cdot 2\cancel{a}b} = \frac{a+b}{2b}$
$\frac{x^2+3x}{x} = \frac{\cancel{x}(x+3)}{\cancel{x}} = \frac{x+3}{1} = x+3$	$\frac{2x^2+6x}{2x^2+8x} = \frac{\cancel{2x}(x+3)}{\cancel{2x}(x+4)} = \frac{x+3}{x+4}$
$\frac{x^2+x}{x^3+x} = \frac{\cancel{x}(x+1)}{\cancel{x}(x^2+1)} = \frac{x+1}{x^2+1}$	$\frac{3x^2+9x}{x^2+3x} = \frac{3\cancel{x}\cancel{(x+3)}}{\cancel{x}\cancel{(x+3)}} = \frac{3}{1} = 3$

Book 6, Page 20

Simplify each rational expression. Some are already factored for you.

$\frac{x^2-2x-8}{x^2+x-20} = \frac{\cancel{(x-4)}(x+2)}{(x+5)\cancel{(x-4)}} = \frac{x+2}{x+5}$	$\frac{x+3}{x^2+5x+6} = \frac{\overset{1}{\cancel{x+3}}}{\cancel{(x+3)}(x+2)} = \frac{1}{x+2}$
$\frac{x^2-6x+8}{2x^2-x-6} = \frac{\cancel{(x-2)}(x-4)}{(2x+3)\cancel{(x-2)}} = \frac{x-4}{2x+3}$	$\frac{x-5}{x^2+x-30} = \frac{\overset{1}{\cancel{x-5}}}{(x+6)\cancel{(x-5)}} = \frac{1}{x+6}$
$\frac{x^2+10x+25}{x^2-25} = \frac{\cancel{(x+5)}(x+5)}{\cancel{(x+5)}(x-5)} = \frac{x+5}{x-5}$	$\frac{x^2-3x-10}{x-5} = \frac{\cancel{(x-5)}(x+2)}{\underset{1}{\cancel{x-5}}} = \frac{x+2}{1} = x+2$
$\frac{x^2+6x+8}{x^2+4x+4} = \frac{\cancel{(x+2)}(x+4)}{\cancel{(x+2)}(x+2)} = \frac{x+4}{x+2}$	$\frac{x^2+7x+12}{x+4} = \frac{(x+3)\cancel{(x+4)}}{\underset{1}{\cancel{x+4}}} = \frac{x+3}{1} = x+3$
$\frac{x^2-16}{x^2+8x+16} = \frac{\cancel{(x+4)}(x-4)}{\cancel{(x+4)}(x+4)} = \frac{x-4}{x+4}$	$\frac{x^2-9}{x-3} = \frac{(x+3)\cancel{(x-3)}}{\underset{1}{\cancel{x-3}}} = \frac{x+3}{1} = x+3$
$\frac{x^2+13x+42}{x^2+2x-24} = \frac{\cancel{(x+6)}(x+7)}{\cancel{(x+6)}(x-4)} = \frac{x+7}{x-4}$	$\frac{x^2-14x+40}{x^2-6x-40} = \frac{(x-4)\cancel{(x-10)}}{\cancel{(x-10)}(x+4)} = \frac{x-4}{x+4}$
$\frac{x^2-16}{x^2+11x+28} = \frac{\cancel{(x+4)}(x-4)}{(x+7)\cancel{(x+4)}} = \frac{x-4}{x+7}$	$\frac{\overset{1}{\cancel{x^2-4x-60}}}{\underset{1}{\cancel{x^2-4x-60}}} = \frac{1}{1} = 1$
$\frac{4x^2-25}{2x^2-3x-5} = \frac{(2x+5)\cancel{(2x-5)}}{\cancel{(2x-5)}(x+1)} = \frac{2x+5}{x+1}$	$\frac{3x+1}{3x^2-14x-5} = \frac{\overset{1}{\cancel{3x+1}}}{\cancel{(3x+1)}(x-5)} = \frac{1}{x-5}$

Book 6, Page 21

Simplify each fraction.

$\frac{a^2+2a}{a^2+3a+2} = \frac{a\cancel{(a+2)}}{\cancel{(a+2)}(a+1)} = \frac{a}{a+1}$	$\frac{3x-3}{x^2+2x-3} = \frac{3\cancel{(x-1)}}{\cancel{(x-1)}(x+3)} = \frac{3}{x+3}$
$\frac{x^2-11x-12}{x^2-12x} = \frac{\cancel{(x-12)}(x+1)}{x\cancel{(x-12)}} = \frac{x+1}{x}$	$\frac{x^2+5x}{x^2+8x+15} = \frac{x\cancel{(x+5)}}{(x+3)\cancel{(x+5)}} = \frac{x}{x+3}$
$\frac{2x^2-14x}{x^2-10x+21} = \frac{2x\cancel{(x-7)}}{\cancel{(x-7)}(x-3)} = \frac{2x}{x-3}$	$\frac{5a^2+10a}{a^2-7a-18} = \frac{5a\cancel{(a+2)}}{(a-9)\cancel{(a+2)}} = \frac{5a}{a-9}$
$\frac{(x+2)^2}{x^2-4} = \frac{\cancel{(x+2)}(x+2)}{\cancel{(x+2)}(x-2)} = \frac{x+2}{x-2}$	$\frac{(x+6)^2}{x^2-36} = \frac{\cancel{(x+6)}(x+6)}{\cancel{(x+6)}(x-6)} = \frac{x+6}{x-6}$
$\frac{(x-5)^2}{x^2-11x+30} = \frac{\cancel{(x-5)}(x-5)}{\cancel{(x-5)}(x-6)} = \frac{x-5}{x-6}$	$\frac{2x^2-3x-20}{(2x+5)^2} = \frac{\cancel{(2x+5)}(x-4)}{\cancel{(2x+5)}(2x+5)} = \frac{x-4}{2x+5}$
$\frac{4x-6}{(2x-3)^2} = \frac{2\cancel{(2x-3)}}{(2x-3)\cancel{(2x-3)}} = \frac{2}{2x-3}$	$\frac{3a^2-12a}{(a-4)^2} = \frac{3a\cancel{(a-4)}}{(a-4)\cancel{(a-4)}} = \frac{3a}{a-4}$

To completely simplify each fraction below you have to factor more than once.

$$\frac{3a^2+15a}{a^3-25a} = \frac{3\cancel{a}(a+5)}{\cancel{a}(a^2-25)} = \frac{3\cancel{(a+5)}}{\cancel{(a+5)}(a-5)} = \frac{3}{a-5}$$

$$\frac{4a-12}{2a^2-18} = \frac{4(a-3)}{2(a^2-9)} = \frac{\overset{2}{\cancel{4}}\cancel{(a-3)}}{\cancel{2}(a+3)\cancel{(a-3)}} = \frac{2}{a+3}$$

$$\frac{2x^2-2x}{2x^3+6x^2-8x} = \frac{\cancel{2x}(x-1)}{\cancel{2x}(x^2+3x-4)} = \frac{\overset{1}{\cancel{x-1}}}{(x+4)\cancel{(x-1)}} = \frac{1}{x+4}$$

$$\frac{x^3-3x^2-10x}{7x^2-35x} = \frac{\cancel{x}(x^2-3x-10)}{7\cancel{x}(x-5)} = \frac{\cancel{(x-5)}(x+2)}{7\cancel{(x-5)}} = \frac{x+2}{7}$$

Book 6, Page 22

Dividing Polynomials

Dividing one polynomial by another can often be done by rewriting the problem as a rational expression and then simplifying it. Here are some examples:

Divide.

$$(2x^2+10x) \div 2x = \frac{2x^2+10x}{2x} = \frac{\overset{1}{\cancel{2x}}(x+5)}{\underset{1}{\cancel{2x}}} = x+5$$

$$(5a^2-10a) \div 5a = \frac{5a^2-10a}{5a} = \frac{\overset{1}{\cancel{5a}}(a-2)}{\cancel{5a}} = a-2$$

$$(x^2+x-20) \div (x-4) = \frac{x^2+x-20}{x-4} = \frac{(x+5)\overset{1}{\cancel{(x-4)}}}{\underset{1}{\cancel{x-4}}} = x+5$$

$$(3x^2+7x+4) \div (x+1) = \frac{3x^2+7x+4}{x+1} = \frac{(3x+4)\overset{1}{\cancel{(x+1)}}}{\underset{1}{\cancel{x+1}}} = 3x+4$$

$$(x^3-16x) \div (x+4) = \frac{x^3-16x}{x+4} = \frac{x(x^2-16)}{x+4} = \frac{x\overset{1}{\cancel{(x+4)}}(x-4)}{\underset{1}{\cancel{x+4}}} = x(x-4)$$

If the trinomial in the last problem had been $3x^2+7x+6$, we couldn't have factored it. Then this method of dividing wouldn't have worked and we would have had to use long division. Long division of polynomials is a lot like long division of whole numbers.

First you make sure that the terms are arranged in order by their exponents, with the largest exponent first.

$$x+1\overline{)3x^2+7x+6}$$

Then you divide the first term of the divisor (x) into the first term of the dividend ($3x^2$) and write the result ($3x$) over the term it matches.

$$\begin{array}{r} 3x \\ x+1\overline{)3x^2+7x+6} \end{array}$$

Next you multiply the result by the whole divisor, write the answer under the dividend, and subtract.

$$\begin{array}{r} 3x \\ x+1\overline{)3x^2+7x+6} \\ \underline{3x^2+3x} \\ 4x \end{array}$$

Bring down the next term and repeat the last two steps until you are finished. If there is a remainder, write it over the divisor and add this to the quotient.

$$\begin{array}{r} 3x+4+\frac{2}{x+1} \\ x+1\overline{)3x^2+7x+6} \\ \underline{3x^2+3x} \\ 4x+6 \\ \underline{4x+4} \\ 2 \end{array}$$

Key to Algebra – ANSWERS

Book 6, Page 23

Finish each division problem. Check each answer with your teacher before doing the next problem. Be sure to ask for help if you need it.

$x+2\,\overline{)\,x^2+9x+18}$ = $x+7+\frac{4}{x+2}$

x^2+2x; $7x+18$; $7x+14$; 4

$x-3\,\overline{)\,2x^2+11x-4}$ = $2x+17+\frac{47}{x-3}$

$2x^2-6x$; $17x-4$; $17x-51$; 47

(Remember, to subtract a polynomial, we add its opposite.)

$x-4\,\overline{)\,3x^2-10x+17}$ = $3x+2+\frac{25}{x-4}$

$3x^2-12x$; $2x+17$; $2x-8$; 25

$2x+3\,\overline{)\,6x^3-x^2-7x+12}$ = $3x^2-5x+4$

$6x^3+9x^2$; $-10x^2-7x$; $-10x^2-15x$; $8x+12$; $8x+12$; 0

Here are some division problems for you to do by yourself.

$(x^2+7x+19)\div(x+4)$ = $x+3+\frac{7}{x+4}$

x^2+4x; $3x+19$; $3x+12$; 7

$(5x^2+x+30)\div(5x+6)$ = $x-1+\frac{36}{5x+6}$

$5x^2+6x$; $-5x+30$; $-5x-6$; 36

$(2x^2+3x+11)\div(x-2)$ = $2x+7+\frac{25}{x-2}$

$2x^2-4x$; $7x+11$; $7x-14$; 25

$(3x^3+5x^2-2x-4)\div(x+1)$ = $3x^2+2x-4$

$3x^3+3x^2$; $2x^2-2x$; $2x^2+2x$; $-4x-4$; $-4x-4$; 0

Book 6, Page 24

Rewriting Fractions in Simplest Form

Here is how Sandy and Terry simplified $\frac{8}{12}$:

Sandy: $\frac{8}{12} = \frac{4}{6}$ (2 goes into both 8 and 12. 2 into 8 is 4. 2 into 12 is 6.)

Terry: $\frac{8}{12} = \frac{2}{3}$ (4 is the biggest number that goes into 8 and 12. 4 into 8 is 2. 4 into 12 is 3.)

Both Sandy and Terry simplified $\frac{8}{12}$, but only Terry rewrote it in **simplest form.** That's because Terry factored out the *biggest* number that goes into 8 and 12, while Sandy factored out a smaller number.

Here's how Sandy finished simplifying $\frac{8}{12}$:

$\frac{8}{12} = \frac{4}{6} = \frac{2}{3}$ (2 goes into both 4 and 6. 2 into 4 is 2. 2 into 6 is 3.)

After simplifying $\frac{8}{12}$ as much as possible, both Sandy and Terry ended up with $\frac{2}{3}$. You can see that Terry's way was shortest. Terry factored out the *biggest* number that goes into the top and bottom of the fraction, so it took only one step to find the simplest form.

Rewrite each fraction in simplest form.

$\frac{20}{12} = \frac{5}{3}$ (4 goes into both 20 and 12. 4 into 20 is 5. 4 into 12 is 3.)		$\frac{30}{20} = \frac{3}{2}$	$\frac{8}{20} = \frac{2}{5}$
$\frac{18}{24} = \frac{3}{4}$	$\frac{16}{20} = \frac{4}{5}$	$\frac{-16}{20} = \frac{-4}{5}$	$\frac{12}{18} = \frac{2}{3}$
$\frac{15}{60} = \frac{1}{4}$	$\frac{-24}{36} = \frac{-2}{3}$	$\frac{60}{40} = \frac{3}{2}$	$\frac{30}{90} = \frac{1}{3}$
$\frac{-15}{60} = \frac{-1}{4}$	$\frac{32}{8} = \frac{4}{1} = 4$	$\frac{-50}{10} = \frac{-5}{1} = -5$	$\frac{48}{6} = \frac{8}{1} = 8$

Book 6, Page 25

A rational expression is in simplest form when we have canceled all possible factors from the numerator and denominator. Write each fraction in simplest form.

$\frac{27a}{36a^3} = \frac{3}{4a^2}$ (9 is the biggest number that goes into 27 and 36.)	$\frac{48x^2y}{32x} = \frac{3xy}{2}$	$\frac{24x^2}{32x^5} = \frac{3}{4x^3}$
$\frac{6ab}{20ab^2} = \frac{3}{10b}$	$\frac{-45xy}{30x^2y} = \frac{-3}{2x}$	$\frac{36n^2}{48mn} = \frac{3n}{4m}$

$\frac{15x+30}{20x} = \frac{15(x+2)}{20x} = \frac{3(x+2)}{4x}$

$\frac{20a-25}{35a} = \frac{5(4a-5)}{35a} = \frac{4a-5}{7a}$

$\frac{12x+30}{18x} = \frac{6(2x+5)}{6\cdot 3x} = \frac{2x+5}{3x}$

$\frac{16x+8}{28x^2} = \frac{8(2x+1)}{28x^2} = \frac{2(2x+1)}{7x^2}$

$\frac{8x+12}{12x+16} = \frac{4(2x+3)}{4(3x+4)} = \frac{2x+3}{3x+4}$

$\frac{9x+45}{18x-36} = \frac{9(x+5)}{18(x-2)} = \frac{x+5}{2(x-2)}$

$\frac{2a^2+4a}{6a^3-6a^2} = \frac{2a(a+2)}{6a^2(a-1)} = \frac{a+2}{3a(a-1)}$

$\frac{6x^3y-24xy}{4x^3-8x^2} = \frac{6xy(x^2-4)}{4x^2(x-2)} = \frac{6xy(x+2)(x-2)}{4x^2(x-2)} = \frac{3y(x+2)}{2x}$

$\frac{4x^2+4x-8}{10x^2-40x+30} = \frac{4(x^2+x-2)}{10(x^2-4x+3)} = \frac{2(x+2)(x-1)}{5(x-3)(x-1)} = \frac{2(x+2)}{5(x-3)}$

$\frac{6x^3+18x^2-60x}{9x^2y-36xy+36y} = \frac{6x(x^2+3x-10)}{9y(x^2-4x+4)} = \frac{6x(x+5)(x-2)}{9y(x-2)(x-2)} = \frac{2x(x+5)}{3y(x-2)}$

Book 6, Page 26

Simplifying Multiplication Problems

When we are multiplying fractions it's usually easier to simplify *before* we multiply rather than after. That way we don't have to deal with such big numbers and complicated polynomials. Here's how Sandy and Terry did the same problem:

Sandy simplified *after* multiplying: $\frac{7}{12}\cdot\frac{17}{21} = \frac{119}{252} = \frac{17}{36}$

(I'll have to try lots of numbers to see if this can be simplified. 2 goes into 252 but not 119. 3 goes into 252 but not 119. 4 goes into 252 but not 119. 5 doesn't go into either. 6 goes into 252 but not 119. 7 goes into both. Finally!)

Terry simplified *before* multiplying: $\frac{7}{12}\cdot\frac{17}{21} = \frac{17}{36}$

Both Sandy and Terry got the same answer, but Sandy wasted a lot of time simplifying $\frac{119}{252}$. Terry's way was much easier. (Notice that Terry canceled across from the top of one fraction to the bottom of the other. This is OK in multiplication problems, because all the numerators and all the denominators are going to be multiplied together anyway.)

Multiply each pair of fractions. Simplify *before* you multiply.

$\frac{-7}{25}\cdot\frac{10}{21} = \frac{-2}{15}$ (7 goes into −7 and 21.) (5 goes into 10 and 25.)	$\frac{4}{7}\cdot\frac{5}{6} = \frac{10}{21}$	$\frac{8}{3}\cdot\frac{9}{10} = \frac{12}{5}$
$\frac{-12}{25}\cdot\frac{5}{6} = \frac{-2}{5}$	$\frac{3}{16}\cdot\frac{-2}{9} = \frac{-1}{24}$	$\frac{-1}{4}\cdot\frac{-4}{5} = \frac{1}{5}$
$\frac{x}{4}\cdot\frac{3}{x} = \frac{3}{4}$	$\frac{3x}{2}\cdot\frac{4}{x} = \frac{6}{1} = 6$	$\frac{y^2}{6}\cdot\frac{4}{y} = \frac{2y}{3}$
$\frac{a}{b}\cdot\frac{b^2}{a} = \frac{b}{1} = b$	$\frac{x}{y^3}\cdot\frac{y}{x^2} = \frac{1}{xy^2}$	$\frac{2c}{5d}\cdot\frac{10}{c^2} = \frac{4}{cd}$
$\frac{5xy}{4}\cdot\frac{x^2}{y^2} = \frac{5x^3}{4y}$	$\frac{5x^3}{24y^5}\cdot\frac{16y^2}{25x} = \frac{2x^2}{15y^3}$	$\frac{8a^2}{15b^4}\cdot\frac{3b^2}{16a^4} = \frac{1}{10a^2b^2}$

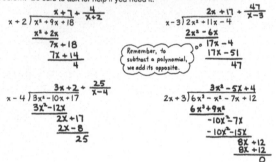

Key to Algebra – ANSWERS

Book 6, Page 27

When the numerator and denominator of a fraction are polynomials, it makes even more sense to simplify before we multiply. The problem below is simplified *after* multiplying.

(To simplify I have to factor.)

$$\frac{x+2}{24x}\cdot\frac{3x}{x^2-4}=\frac{3x^2+6x}{24x^3-96x}=\frac{3x(x+2)}{24x(x^2-4)}=\frac{\cancel{3x}\cancel{(x+2)}}{\cancel{24x}\cancel{(x+2)}(x-2)}=\frac{1}{8(x-2)}$$

(This can be factored again.)

You can see that the effort spent in multiplying was wasted, because when we factored we got back what we started with. Here is the same problem simplified *before* multiplying.

$$\frac{x+2}{\cancel{24x}}\cdot\frac{\cancel{3x}}{x^2-4}=\frac{\cancel{x+2}}{8}\cdot\frac{1}{\cancel{(x+2)}(x-2)}=\frac{1}{8(x-2)}$$

Be smart. Simplify each problem *before* you multiply.

$$\frac{3x+6}{5x}\cdot\frac{x+4}{x^2+2x}=\frac{3\cancel{(x+2)}}{5x}\cdot\frac{x+4}{x\cancel{(x+2)}}=\frac{3(x+4)}{5x^2}$$

$$\frac{8a}{a^2-16}\cdot\frac{a+4}{4}=\frac{2a}{\cancel{(a+4)}(a-4)}\cdot\frac{\cancel{a+4}}{1}=\frac{2a}{a-4}$$

$$\frac{x^2-1}{3}\cdot\frac{2}{x^2-x}=\frac{(x+1)\cancel{(x-1)}}{3}\cdot\frac{2}{x\cancel{(x-1)}}=\frac{2(x+1)}{3x}$$

$$\frac{x^2-x-12}{x^2}\cdot\frac{x}{x-4}=\frac{\cancel{(x-4)}(x+3)}{\cancel{x^2}}\cdot\frac{\cancel{x}}{\cancel{x-4}}=\frac{x+3}{x}$$

$$\frac{x^2+5x+6}{x^2-1}\cdot\frac{x+1}{x+2}=\frac{\cancel{(x+2)}(x+3)}{\cancel{(x+1)}(x-1)}\cdot\frac{\cancel{x+1}}{\cancel{x+2}}=\frac{x+3}{x-1}$$

$$\frac{x^2+6x+8}{x^2-4x+3}\cdot\frac{x^2-5x+4}{5x+10}=\frac{\cancel{(x+2)}(x+4)}{\cancel{(x-1)}(x-3)}\cdot\frac{\cancel{(x-1)}(x-4)}{5\cancel{(x+2)}}=\frac{(x+4)(x-4)}{5(x-3)}$$

Book 6, Page 28

Do you think multiplication of rational expressions is commutative and associative? Make a guess after you do these problems. (Remember to simplify before you multiply.)

$$\frac{2}{3}\cdot\frac{-3}{8}=\frac{-1}{4}\qquad \frac{-3}{8}\cdot\frac{2}{3}=\frac{-1}{4}$$

$$\frac{2}{14}\cdot\frac{-5}{7}=\frac{-10}{1}=-10\qquad \frac{-5}{7}\cdot\frac{14}{1}=\frac{-10}{1}=-10$$

$$\frac{x^3}{3y}\cdot\frac{y}{4x}=\frac{x}{12}\qquad \frac{y}{4x}\cdot\frac{x^2}{3y}=\frac{x}{12}$$

$$\frac{x+4}{x-3}\cdot\frac{2x-6}{3x+12}=\frac{\cancel{x+4}}{\cancel{x-3}}\cdot\frac{2\cancel{(x-3)}}{3\cancel{(x+4)}}=\frac{2}{3}\qquad \frac{2x-6}{3x+12}\cdot\frac{x+4}{x-3}=\frac{2(x-3)}{3(x+4)}\cdot\frac{x+4}{x-3}=\frac{2}{3}$$

$$\frac{x^2-64}{x+2}\cdot\frac{x^2+4x+4}{x^2+7x-8}=\frac{\cancel{(x+8)}(x-8)}{\cancel{x+2}}\cdot\frac{\cancel{(x+2)}(x+2)}{\cancel{(x+8)}(x-1)}=\frac{(x-8)(x+2)}{x-1}$$

$$\frac{x^2+4x+4}{x^2+7x-8}\cdot\frac{x^2-64}{x+2}=\frac{\cancel{(x+2)}(x+2)}{\cancel{(x+8)}(x-1)}\cdot\frac{\cancel{(x+8)}(x-8)}{\cancel{x+2}}=\frac{(x+2)(x-8)}{x-1}$$

$$\left(\frac{1}{3}\cdot\frac{4}{5}\right)\cdot\frac{9}{2}=\frac{4}{15}\cdot\frac{9}{2}=\frac{6}{5}\qquad \left(\frac{1}{x}\cdot\frac{3}{y}\right)\cdot\frac{y}{x}=\frac{3}{xy}\cdot\frac{y}{x}=\frac{3}{x^2}$$

$$\frac{1}{3}\left(\frac{4}{5}\cdot\frac{9}{2}\right)=\frac{1}{3}\cdot\frac{18}{5}=\frac{6}{5}\qquad \frac{1}{x}\left(\frac{3}{y}\cdot\frac{y}{x}\right)=\frac{1}{x}\cdot\frac{3y}{xy}=\frac{3}{x^2}$$

$$\left(\frac{3}{8}\cdot\frac{2}{9}\right)\cdot\frac{5}{4}=\frac{1}{12}\cdot\frac{5}{4}=\frac{5}{48}\qquad 5\left(\frac{1}{2}\cdot\frac{x}{9}\right)=5\cdot\frac{x}{18}=\frac{5x}{18}$$

$$\frac{3}{8}\left(\frac{2}{9}\cdot\frac{5}{4}\right)=\frac{3}{8}\cdot\frac{5}{18}=\frac{5}{48}\qquad \left(5\cdot\frac{1}{2}\right)\cdot\frac{x}{9}=\frac{5}{2}\cdot\frac{x}{9}=\frac{5x}{18}$$

$$\left(\frac{x+3}{x-2}\cdot\frac{x-3}{x+2}\right)\frac{1}{x}=\frac{x^2-9}{x^2-4}\cdot\frac{1}{x}=\frac{x^2-9}{x^3-4x}$$

$$\frac{x+3}{x-2}\left(\frac{x-3}{x+2}\cdot\frac{1}{x}\right)=\frac{x+3}{x-2}\cdot\frac{x-3}{x^2+2x}=\frac{x^2-9}{x^3+2x^2-2x^2-4x}=\frac{x^2-9}{x^3-4x}$$

Book 6, Page 29

If you guessed that multiplication of rational expressions is commutative and associative you were right. (A few examples aren't enough to prove this, but it *is* true.) The Distributive Principle works for rational expressions, too. Use it to do these problems.

$$12\left(\frac{1}{3}+\frac{1}{2}\right)=12\cdot\frac{1}{3}+12\cdot\frac{1}{2}=4+6=10$$

$$20\left(\frac{1}{5}+\frac{3}{10}\right)=20\cdot\frac{1}{5}+20\cdot\frac{3}{10}=4+6=10$$

$$18\left(\frac{1}{2}-\frac{2}{9}\right)=18\cdot\frac{1}{2}-18\cdot\frac{2}{9}=9-4=5$$

$$24\left(\frac{x}{3}+\frac{x}{8}\right)=24\cdot\frac{x}{3}+24\cdot\frac{x}{8}=8x+3x=11x$$

$$18\left(\frac{x}{9}+\frac{4x}{6}\right)=18\cdot\frac{x}{9}+18\cdot\frac{4x}{6}=2x+12x=14x$$

$$x\left(\frac{3}{x}+\frac{5}{x}\right)=x\cdot\frac{3}{x}+x\cdot\frac{5}{x}=3+5=8$$

$$x^2\left(\frac{7}{x}-\frac{4}{x}\right)=x^2\cdot\frac{7}{x}-x^2\cdot\frac{4}{x}=7x-4x=3x$$

$$12x\left(\frac{1}{2x}+\frac{3}{4x}\right)=12x\cdot\frac{1}{2x}+12x\cdot\frac{3}{4x}=6+9=15$$

$$20x\left(\frac{1}{10x}-\frac{1}{5x}\right)=20x\cdot\frac{1}{10x}-20x\cdot\frac{1}{5x}=2-4=-2$$

$$x^2\left(\frac{4}{x}+\frac{3}{x^2}\right)=x^2\cdot\frac{4}{x}+x^2\cdot\frac{3}{x^2}=4x+3$$

$$x^2\left(\frac{2}{x^2}+\frac{6}{x}\right)=x^2\cdot\frac{2}{x^2}+x^2\cdot\frac{6}{x}=2+6x$$

$$5x^2\left(\frac{3}{5}+\frac{2}{x^2}\right)=5x^2\cdot\frac{3}{5}+5x^2\cdot\frac{2}{x^2}=3x^2+10$$

$$12a\left(\frac{5}{3a}-1\right)=12a\cdot\frac{5}{3a}-12a\cdot1=20-12a$$

$$4a^2\left(\frac{4}{a}+\frac{1}{2a^2}\right)=4a^2\cdot\frac{4}{a}+4a^2\cdot\frac{1}{2a^2}=16a+2$$

$$abc\left(\frac{1}{ab}+\frac{1}{bc}\right)=abc\cdot\frac{1}{ab}+abc\cdot\frac{1}{bc}=c+a$$

$$100x\left(\frac{1}{20}-\frac{3}{x}\right)=100x\cdot\frac{1}{20}-100x\cdot\frac{3}{x}=5x-300$$

Book 6, Page 30

Reciprocals

Every number except 0 has a **reciprocal**. When we multiply a number by its reciprocal we always get 1.

"The reciprocal of $\frac{2}{3}$ is $\frac{3}{2}$." $\quad \frac{2}{3}\cdot\frac{3}{2}=1$

"The reciprocal of $\frac{1}{8}$ is $\frac{8}{1}$." $\quad \frac{1}{8}\cdot\frac{8}{1}=1$

"$\frac{-4}{3}$ is the reciprocal of $\frac{3}{4}$." $\quad \frac{-3}{4}\cdot\frac{4}{-3}=1$

"$\frac{9}{x+1}$ is the reciprocal of $\frac{x+1}{9}$." $\quad \frac{x+1}{9}\cdot\frac{9}{x+1}=1$

Do you see a rule for finding a reciprocal?

Find the reciprocal of each fraction.

Fraction	$\frac{3}{8}$	$\frac{1}{6}$	$\frac{7}{2}$	$\frac{2}{7}$	$\frac{-5}{3}$	$\frac{x}{4}$	$\frac{6}{y}$	$\frac{5a}{2}$	$\frac{1}{3x}$	$\frac{x-4}{3}$	$\frac{x+1}{x-1}$
Reciprocal	$\frac{8}{3}$	$\frac{6}{1}$	$\frac{2}{7}$	$\frac{7}{2}$	$\frac{3}{-5}$	$\frac{4}{x}$	$\frac{y}{6}$	$\frac{2}{5a}$	$\frac{3x}{1}$	$\frac{3}{x-4}$	$\frac{x-1}{x+1}$

Complete each sentence below.

$10=\frac{10}{1}$, so the reciprocal of 10 is $\frac{1}{10}$.	$y=\frac{y}{1}$, so the reciprocal of y is $\frac{1}{y}$.
$-2=\frac{-2}{1}$, so the reciprocal of -2 is $\frac{1}{-2}$.	$-a=\frac{-a}{1}$, so the reciprocal of $-a$ is $\frac{1}{-a}$.
$4\frac{2}{3}=\frac{14}{3}$, so the reciprocal of $4\frac{2}{3}$ is $\frac{3}{14}$.	$x+3=\frac{x+3}{1}$, so the reciprocal of $x+3$ is $\frac{1}{x+3}$.
$0.5=\frac{5}{10}$, so the reciprocal of 0.5 is $\frac{10}{5}$.	$5xy=\frac{5xy}{1}$, so the reciprocal of $5xy$ is $\frac{1}{5xy}$.
$1.3=\frac{13}{10}$, so the reciprocal of 1.3 is $\frac{10}{13}$.	$x^2-4=\frac{x^2-4}{1}$, so the reciprocal of x^2-4 is $\frac{1}{x^2-4}$.

To find the reciprocal of a number, write the number as a fraction and then invert it (turn it upside down).

Key to Algebra – ANSWERS

Book 6, Page 31

Dividing Fractions

When we divide an integer by another integer we get the same answer as when we *multiply* the first integer by the *reciprocal* of the second integer. Here are two examples:

$$\begin{cases} 12 \div 4 = 3 \\ 12 \cdot \frac{1}{4} = \frac{12}{1} \cdot \frac{1}{4} = 3 \end{cases} \qquad \begin{cases} -18 \div 2 = -9 \\ -18 \cdot \frac{1}{2} = \frac{-18}{1} \cdot \frac{1}{2} = -9 \end{cases}$$

The same thing is true when we divide a fraction by a fraction. We can get the answer by multiplying the first fraction by the reciprocal of the second fraction. So when we divide fractions we first have to rewrite the problem:

The division sign gets changed to a multiplication sign.
And the second fraction gets replaced by its reciprocal.

Then we multiply to get the answer.

Do each division problem by changing it to multiplication.

$\frac{6}{7} \div \frac{5}{8} = \frac{6}{7} \cdot \frac{8}{5} = \frac{48}{35}$	$\frac{1}{3} \div \frac{1}{2} = \frac{1}{3} \cdot \frac{2}{1} = \frac{2}{3}$
$\frac{-3}{10} \div \frac{2}{7} = \frac{-3}{10} \cdot \frac{7}{2} = \frac{-21}{20}$	$\frac{5}{2} \div 3\frac{1}{3} = \frac{5}{2} \div \frac{10}{3} = \frac{5}{2} \cdot \frac{3}{10} = \frac{3}{4}$
$1\frac{3}{4} \div \frac{5}{2} = \frac{7}{4} \div \frac{5}{2} = \frac{7}{4} \cdot \frac{2}{5} = \frac{7}{10}$	$8 \div \frac{2}{5} = \frac{8}{1} \cdot \frac{5}{2} = \frac{20}{1} = 20$
$\frac{3x}{5} \div \frac{4}{x} = \frac{3x}{5} \cdot \frac{x}{4} = \frac{3x^2}{20}$	$\frac{8a}{5} \div \frac{b}{a} = \frac{8a}{5} \cdot \frac{a}{b} = \frac{8a^2}{5b}$
$\frac{x^2}{y} \div \frac{x^2}{y} = \frac{x^2}{y} \cdot \frac{y}{x^2} = \frac{1}{1} = 1$	$\frac{2a}{5b} \div 3b^2 = \frac{2a}{5b} \cdot \frac{1}{3b^2} = \frac{2a}{15b^3}$
$\frac{a^2b^3}{4} \div \frac{7}{ab} = \frac{a^2b^3}{4} \cdot \frac{ab}{7} = \frac{a^3b^4}{28}$	$\frac{x}{3a} \div \frac{y}{4b} = \frac{x}{3a} \cdot \frac{4b}{y} = \frac{4bx}{3ay}$

$\frac{3}{x} \div \frac{x+1}{x-2} = \frac{3}{x} \cdot \frac{x-2}{x+1} = \frac{3(x-2)}{x(x+1)} = \frac{3x-6}{x^2+x}$

$(a-5) \div \frac{a+2}{8} = \frac{a-5}{1} \cdot \frac{8}{a+2} = \frac{8(a-5)}{a+2} = \frac{8a-40}{a+2}$

$\frac{x+6}{x-2} \div \frac{x+2}{x-6} = \frac{x+6}{x-2} \cdot \frac{x-6}{x+2} = \frac{(x+6)(x-6)}{(x-2)(x+2)} = \frac{x^2-36}{x^2-4}$

Book 6, Page 32

Each division problem below can be simplified. First rewrite the problem as a multiplication problem. Then simplify before multiplying.

$\frac{3}{2} \div \frac{1}{4} = \frac{3}{2} \cdot \frac{4}{1} = \frac{6}{1} = 6$	$\frac{-5}{6} \div \frac{35}{36} = \frac{-5}{6} \cdot \frac{36}{35} = \frac{-6}{7}$
$\frac{8}{9} \div 24 = \frac{8}{9} \cdot \frac{1}{24} = \frac{1}{27}$	$15 \div \frac{5}{3} = \frac{15}{1} \cdot \frac{3}{5} = \frac{9}{1} = 9$
$\frac{3}{2x} \div \frac{6}{5} = \frac{3}{2x} \cdot \frac{5}{6} = \frac{5}{4x}$	$\frac{4x}{3} \div \frac{x^2}{6} = \frac{4x}{3} \cdot \frac{6}{x^2} = \frac{8}{x}$
$\frac{-9a}{4b} \div \frac{a^2}{b^2} = \frac{-9a}{4b} \cdot \frac{b^2}{a^2} = \frac{-9b}{4a}$	$\frac{7y}{x^2} \div \frac{4y}{3x^2} = \frac{7y}{x^2} \cdot \frac{3x^2}{4y} = \frac{21}{4}$
$\frac{x^2y^2}{6} \div 2xy = \frac{x^2y^2}{6} \cdot \frac{1}{2xy} = \frac{xy}{12}$	$4ab^2 \div \frac{12a}{7b^2} = \frac{4ab^2}{1} \cdot \frac{7b^2}{12a} = \frac{7b^4}{3}$

$\frac{3x+12}{x^2} \div \frac{x+4}{x} = \frac{3x+12}{x^2} \cdot \frac{x}{x+4} = \frac{3(x+4)}{x^2} \cdot \frac{x}{x+4} = \frac{3}{x}$

$\frac{2a+6}{a^3} \div \frac{a+3}{a} = \frac{2a+6}{a^3} \cdot \frac{a}{a+3} = \frac{2(a+3)}{a^3} \cdot \frac{a}{a+3} = \frac{2}{a^2}$

$\frac{10x}{3x-9} \div \frac{10x^2}{3x^2-9x} = \frac{10x}{3x-9} \cdot \frac{3x^2-9x}{10x^2} = \frac{10x}{3(x-3)} \cdot \frac{3x(x-3)}{10x^2} = \frac{1}{1} = 1$

$\frac{x^2+3x-10}{x^2+3x} \div \frac{x^2-4x+4}{4x+12} = \frac{(x+5)(x-2)}{x(x+3)} \cdot \frac{4(x+3)}{(x-2)(x-2)} = \frac{4(x+5)}{x(x-2)}$

Book 6, Page 33

Some of these are multiplication problems and some are division problems.
Be sure you look closely before you start each one. Always simplify if you can.

$\frac{9}{14} \div \frac{15}{8} = \frac{9}{14} \cdot \frac{8}{15} = \frac{12}{35}$	$\frac{-3}{8} \cdot \frac{4}{-9} = \frac{1}{6}$	$\frac{20}{21} \div \frac{7}{10} = \frac{20}{21} \cdot \frac{10}{7} = \frac{200}{147}$
$\frac{28x}{33y} \div \frac{8x}{3y} = \frac{28x}{33y} \cdot \frac{3y}{8x} = \frac{7}{22}$	$\frac{14n^2}{9t} \cdot \frac{21t^2}{2n} = \frac{49n^2t}{3}$	$\frac{6a}{7b^2} \div 42ab^2 = \frac{6a}{7b^2} \cdot \frac{1}{42ab^2} = \frac{1}{49b^4}$

$\frac{2}{x+1} \cdot \frac{x-5}{3} = \frac{2(x-5)}{3(x+1)}$	$\frac{2}{x+1} \div \frac{x-5}{3} = \frac{2}{x+1} \cdot \frac{3}{x-5} = \frac{6}{(x+1)(x-5)}$
$\frac{5}{8} \cdot \frac{5x^2}{8x-1} = \frac{25x^2}{8(8x-1)}$	$\frac{5}{8} \div \frac{5x^2}{8x-1} = \frac{5}{8} \cdot \frac{8x-1}{5x^2} = \frac{8x-1}{8x^2}$
$(x-10) \cdot \frac{x-10}{x+10} = \frac{(x-10)(x-10)}{x+10}$	$(x-10) \div \frac{x-10}{x+10} = \frac{x-10}{1} \cdot \frac{x+10}{x-10} = x+10$
$\frac{x+3}{6x} \cdot \frac{x-1}{2x(x+3)} = \frac{x-1}{12x^2}$	$\frac{x+3}{6x} \div \frac{x-1}{2x(x+3)} = \frac{x+3}{6x} \cdot \frac{2x(x+3)}{x-1} = \frac{(x+3)(x+3)}{3(x-1)}$

$\frac{x^2+8x}{9x} \div \frac{x^2-64}{3x^2} = \frac{x(x+8)}{9x} \cdot \frac{3x^2}{(x+8)(x-8)} = \frac{x^2}{3(x-8)}$

$\frac{x(x-y)}{y} \cdot \frac{y}{2(x-y)} = \frac{xy}{2}$

$\frac{x^2+5x-14}{x^2-4x-5} \cdot \frac{(x+1)^2}{x^2-49} = \frac{(x+7)(x-2)}{(x-5)(x+1)} \cdot \frac{(x+1)(x+1)}{(x+7)(x-7)} = \frac{(x-2)(x+1)}{(x-5)(x-7)}$

Book 6, Page 34

Each problem on this page has a simple answer. Work carefully and take your time.

$\frac{25a^3b}{12c^5} \cdot \frac{9abc}{35d^4} \div \frac{15a^4b^2}{14c^4d^4} = \frac{25a^3b}{12c^5} \cdot \frac{9abc}{35d^4} \cdot \frac{14c^4d^4}{15a^4b^2} = \frac{1}{2}$

$(3\frac{1}{2})(3\frac{1}{3})(2\frac{2}{5})(4\frac{3}{8})(2\frac{2}{11})(1\frac{4}{7}) = \frac{7}{2} \cdot \frac{10}{3} \cdot \frac{12}{5} \cdot \frac{35}{8} \cdot \frac{24}{11} \cdot \frac{11}{7} = \frac{420}{1} = 420$

$\frac{3x(x+7)}{(x+7)(x-7)} \cdot \frac{x(x-1)}{3x^2(2x-3)} \cdot \frac{(2x+3)(2x-3)}{3(x-1)} = \frac{2x+3}{3(x-7)} = \frac{2x+3}{3x-21}$

(originally: $\frac{3x^2+21x}{x^2-49} \cdot \frac{x^2-x}{6x^3-9x^2} \cdot \frac{4x^2-9}{3x-3}$)

$\frac{3x^2y^2+9xy^2-30y^2}{4x^3+4x^2-168x} \cdot \left(\frac{5x^2-245}{3x^2y-75y} \div \frac{5x^2y-45xy+70y}{6x^3-66x^2+180x}\right) =$

$\frac{3x^2y^2+9xy^2-30y^2}{4x^3+4x^2-168x} \cdot \left(\frac{5x^2-245}{3x^2y-75y} \cdot \frac{6x^3-66x^2+180x}{5x^2y-45xy+70y}\right) =$

$\frac{3y^2(x^2+3x-10)}{4x(x^2+x-42)} \cdot \left(\frac{5(x^2-49)}{3y(x^2-25)}\right) \cdot \frac{6x(x^2-11x+30)}{5y(x^2-9x+14)} = \frac{3}{2}$

Key to Algebra – ANSWERS

Book 6, Page 35

Answers to Written Work

① Because $\frac{2}{3-3} = \frac{2}{0}$, and we can't divide by zero.

② $x \neq 0, 0$ $\quad x-4 \neq 0, 4$ $\quad 2x \neq 0, 0$

$x+1 \neq 0, -1$ $\quad 2x-7 \neq 0, \frac{7}{2}$

③

Rational number	-1000	0.9	0.03	0.237	$11\frac{1}{2}$	$5\frac{2}{3}$	-2.7
Fraction	$\frac{-1000}{1}$	$\frac{9}{10}$	$\frac{3}{100}$	$\frac{237}{1000}$	$\frac{23}{2}$	$\frac{17}{3}$	$\frac{-27}{10}$
Reciprocal	$\frac{1}{-1000}$	$\frac{10}{9}$	$\frac{100}{3}$	$\frac{1000}{237}$	$\frac{2}{23}$	$\frac{3}{17}$	$\frac{10}{-27}$

④ $\frac{\cancel{4x}^{2}}{\cancel{2x}_{1}} = \frac{2}{1}$ $\qquad \frac{4x-2}{2x} = \frac{\cancel{2}(2x-1)}{\cancel{2}x} = \frac{2x-1}{x}$

$\frac{x+2}{2x}$ already in simplest form

$\frac{x^2+4x}{2x} = \frac{\cancel{x}(x+4)}{2\cancel{x}} = \frac{x+4}{2}$

$\frac{x^2+4}{x+2}$ already in simplest form

$\frac{x^2-4}{x-2} = \frac{(x+2)\cancel{(x-2)}}{\cancel{x-2}} = x+2$

$\frac{x^2-4}{x+2} = \frac{\cancel{(x+2)}(x-2)}{\cancel{x+2}} = x-2$

Book 6, Page 35 (cont.)

Answers to Written Work

⑤ $\frac{3}{8} \cdot \frac{x}{x-1} = \frac{3x}{8(x-1)} = \frac{3x}{8x-8}$ $\qquad \frac{-2}{\cancel{x}} \cdot \frac{\cancel{x}}{x-1} = \frac{-2}{x-1}$

$\frac{x}{x-1} \cdot \frac{x}{x-1} = \frac{x^2}{(x-1)(x-1)} = \frac{x^2}{x^2-2x+1}$

$\frac{\cancel{x-1}}{\cancel{x}} \cdot \frac{\cancel{x}}{\cancel{x-1}} = \frac{1}{1} = 1$

$\frac{x^2-1}{x^2+1} \cdot \frac{x}{x-1} = \frac{(x+1)\cancel{(x-1)}}{x^2+1} \cdot \frac{x}{\cancel{x-1}} = \frac{x(x+1)}{x^2+1} = \frac{x^2+x}{x^2+1}$

$\frac{x+5}{x-2} \cdot \frac{x}{x+1} = \frac{x(x+5)}{(x-2)(x+1)} = \frac{x^2+5x}{x^2-x-2}$

⑥ $\frac{-1}{3} \div \frac{x}{y^2} = \frac{-1}{3} \cdot \frac{y^2}{x} = \frac{-y^2}{3x}$ $\qquad \frac{x}{y^2} \div \frac{x}{y^2} = \frac{\cancel{x}}{\cancel{y^2}} \cdot \frac{\cancel{y^2}}{\cancel{x}} = \frac{1}{1} = 1$

$\frac{y^2}{x} \div \frac{x}{y^2} = \frac{y^2}{x} \cdot \frac{y^2}{x} = \frac{y^4}{x^2}$ $\qquad \frac{2y}{7x} \div \frac{x}{y^2} = \frac{2y}{7x} \cdot \frac{y^2}{x} = \frac{2y^3}{7x^2}$

$\frac{x^2+3x}{5y^2} \div \frac{x}{y^2} = \frac{\cancel{x}(x+3)}{5\cancel{y^2}} \cdot \frac{\cancel{y^2}}{\cancel{x}} = \frac{x+3}{5}$

$\frac{x}{xy^2-5y^2} \div \frac{x}{y^2} = \frac{\cancel{x}}{\cancel{y^2}(x-5)} \cdot \frac{\cancel{y^2}}{\cancel{x}} = \frac{1}{x-5}$

⑦ Multiplying by $\frac{1}{7}$ gives the same answer as dividing by 7.

Dividing by 4 gives the same answer as multiplying by $\frac{1}{4}$.

Book 6, Page 35 (cont.)

Answers to Written Work

⑧ $\frac{1}{7}(3x-2) = \frac{3x-2}{7}$ $\qquad \frac{1}{7}a^2b^2 = \frac{a^2b^2}{7}$

$\frac{xy^3}{4} = \frac{1}{4}xy^3$ $\qquad \frac{x^2-1}{4} = \frac{1}{4}(x^2-1)$

⑨ $\frac{3x+15}{3} = \frac{1}{3}(3x+15) = \frac{1}{\cancel{3}} \cdot \cancel{3}x + \frac{1}{\cancel{3}} \cdot \cancel{15}^{5} = x+5$

$\frac{2y-20}{-4} = \frac{1}{-4}(2y-20) = \frac{-1}{4}(2y+-20) = \frac{-1}{\cancel{4}_2} \cdot \cancel{2}y + \frac{-1}{\cancel{4}} \cdot \cancel{-20}^{-5}$

$= \frac{-1y}{2} + 5$

$\frac{6a+27}{-3} = \frac{1}{-3}(6a+27) = \frac{-1}{\cancel{3}} \cdot \cancel{6}^{2}a + \frac{-1}{\cancel{3}} \cdot \cancel{27}^{9} = -2a+-9 = -2a-9$

$\frac{5x^2+40x}{5x} = \frac{1}{5x}(5x^2+40x) = \frac{1}{\cancel{5x}} \cdot \cancel{5x^2}^{x} + \frac{1}{\cancel{5x}} \cdot \cancel{40x}^{8} = x+8$

⑩ Yes, there is a Distributive Principle for division. For example, the division problems in #9 could have been done as follows:

$\frac{3x+15}{3} = \frac{\cancel{3}x}{\cancel{3}} + \frac{\cancel{15}^{5}}{\cancel{3}} = x+5$

$\frac{2y-20}{-4} = \frac{2y+-20}{-4} = \frac{2y}{-4} + \frac{-20}{-4} = \frac{y}{-2} + 5$

$\frac{6a+27}{-3} = \frac{6a}{-3} + \frac{27}{-3} = -2a+-9 = -2a-9$

$\frac{5x^2+40x}{5x} = \frac{\cancel{5x^2}^{x}}{\cancel{5x}} + \frac{\cancel{40x}^{8}}{\cancel{5x}} = x+8$

Book 6, Page 36

Practice Test

Circle the expression which is not a rational expression.

$\frac{x}{8}$ $\qquad \frac{x+3}{x-2}$ $\qquad \frac{0}{x^2}$ $\qquad$ (circled) $\frac{x-5}{0}$ $\qquad \frac{-3}{4}$ $\qquad \frac{x^2+2x-5}{x+3}$

Write each fraction in simplest form.

$\frac{3-10}{7-10} = \frac{-7}{-3} = \frac{7}{3}$ $\qquad \frac{3a+24}{5a+40} = \frac{3\cancel{(a+8)}}{5\cancel{(a+8)}} = \frac{3}{5}$

$\frac{\cancel{42}^{3}a}{\cancel{56}_{4}} = \frac{3a}{4}$ $\qquad \frac{x^2-16}{x^2-4x} = \frac{(x+4)\cancel{(x-4)}}{x\cancel{(x-4)}} = \frac{x+4}{x}$

$\frac{\cancel{5x^3}^{x^2}}{\cancel{15x}_{3}y^2} = \frac{x^2}{3y^2}$ $\qquad \frac{15a+35}{12a^2+28a} = \frac{5\cancel{(3a+7)}}{4a\cancel{(3a+7)}} = \frac{5}{4a}$

$\frac{\cancel{-56}^{-7}ab^3}{\cancel{24a^4}_{3a^3}} = \frac{-7b^3}{3a^3}$ $\qquad \frac{x^2+5x-14}{x^2-3x+2} = \frac{(x+7)\cancel{(x-2)}}{\cancel{(x-2)}(x-1)} = \frac{x+7}{x-1}$

Multiply. Make sure your answer is in simplest form.

$\frac{\cancel{-4}^{-1}}{\cancel{3}_{1}} \cdot \frac{\cancel{-3}^{-1}}{\cancel{4}_{1}} = \frac{1}{1} = 1$ $\qquad \frac{x}{1} \cdot \frac{-x}{y} = \frac{-x^2}{y}$

$\left(\frac{3x}{y}\right)^2 = \frac{3x}{y} \cdot \frac{3x}{y} = \frac{9x^2}{y^2}$ $\qquad \left(\frac{x-5}{x+6}\right)^2 = \frac{(x-5)}{(x+6)} \cdot \frac{(x-5)}{(x+6)} = \frac{(x-5)^2}{(x+6)^2}$

$\frac{5x^3}{\cancel{2}y^2} \cdot \frac{\cancel{14x^2}^{7x}}{3xy} = \frac{35x^4}{3y^3}$ $\qquad \frac{\cancel{x-5}}{\cancel{x+3}} \cdot \frac{\cancel{x+3}}{\cancel{x^2-25}_{(x+5)\cancel{(x-5)}}} = \frac{1}{x+5}$

$\frac{x^2-4}{x^2-6x+9} \cdot \frac{x^2-9}{x^2+4x+4} = \frac{\cancel{(x+2)}(x-2)}{\cancel{(x-3)}(x-3)} \cdot \frac{(x+3)\cancel{(x-3)}}{(x+2)\cancel{(x+2)}} = \frac{(x-2)(x+3)}{(x-3)(x+2)}$

Key to Algebra – ANSWERS

Book 6, Page 37

Write the reciprocal of each number or expression.

$\frac{5}{3}$ — $\frac{3}{5}$ $\quad$ $\frac{-1}{x}$ — $\frac{x}{-1}$ $\quad$ $\frac{x+2}{x-6}$ — $\frac{x-6}{x+2}$

x^2 — $\frac{1}{x^2}$ $\quad$ $0.7 = \frac{7}{10}$ — $\frac{10}{7}$ $\quad$ $\frac{x^2+1}{4}$ — $\frac{4}{x^2+1}$

Divide. Make sure your answer is in simplest form.

$$\frac{-3}{14} \div \frac{12}{7} = \frac{-3}{14} \cdot \frac{7}{12} = \frac{-1}{8} \qquad \frac{a}{b} \div \frac{a}{1} = \frac{a}{b} \cdot \frac{1}{a} = \frac{1}{b}$$

$$\frac{x^3y^3}{4} \div \frac{x^2y}{5} = \frac{x^3y^3}{4} \cdot \frac{5}{x^2y} = \frac{5xy^2}{4} \qquad \frac{1}{x} \div \frac{1}{y} = \frac{1}{x} \cdot \frac{y}{1} = \frac{y}{x}$$

$$\frac{5x}{5x-10} \div \frac{3x^2}{4x-8} = \frac{5x}{5x-10} \cdot \frac{4x-8}{3x^2} = \frac{5x}{5(x-2)} \cdot \frac{4(x-2)}{3x^2} = \frac{4}{3x}$$

$$\frac{x^2-81}{x^2-x-20} \div \frac{x^2-8x-9}{x^2+8x+16} = \frac{x^2-81}{x^2-x-20} \cdot \frac{x^2+8x+16}{x^2-8x-9} = \frac{(x+9)(x-9)}{(x-5)(x+4)} \cdot \frac{(x+4)(x+4)}{(x-9)(x+1)} = \frac{(x+9)(x+4)}{(x-5)(x+1)}$$

$$(x^2+9x+14) \div (x+2) = \frac{x^2+9x+14}{x+2} = \frac{(x+7)(x+2)}{x+2} = x+7$$

Divide.

$(2x^2-5x+4) \div (x-3) = 2x+1+\frac{7}{x-3}$

$$\begin{array}{r} 2x^2-5x+4 \\ -(2x^2-6x) \\ \hline x+4 \\ -(x-3) \\ \hline 7 \end{array}$$

$(x^2+5x-8) \div (x-1) = x+6+\frac{-2}{x-1}$

$$\begin{array}{r} x^2+5x-8 \\ -(x^2-x) \\ \hline 6x-8 \\ -(6x-6) \\ \hline -2 \end{array}$$

Book 7, Page 1

Review

In this book you will learn how to add and subtract fractions (both rational numbers and rational expressions). To do this you will have to remember how to add, subtract, multiply and factor polynomials, how to simplify rational expressions and how to find equivalent expressions in higher terms. Test your memory by doing these problems.

Simplify.

$$\frac{3a}{3b} = \frac{a}{b} \qquad \frac{3a+3b}{a+b} = \frac{3(a+b)}{a+b} = 3$$

$$\frac{x-4}{x^2-16} = \frac{x-4}{(x+4)(x-4)} = \frac{1}{x+4} \qquad \frac{2x+8}{x^2+7x+12} = \frac{2(x+4)}{(x+3)(x+4)} = \frac{2}{x+3}$$

Find an equivalent fraction.

$$\frac{x \cdot 5}{y \cdot 5} = \frac{5x}{5y} \qquad \frac{x \cdot xy}{12 \cdot xy} = \frac{x^2y}{12xy} \qquad \frac{3(x-1)}{(x+2)(x-1)} = \frac{3(x-1)}{(x+2)(x-1)}$$

Add.

$(x^2+4x-6)+(2x^2+x-8) = 3x^2+5x-14$

$(x^2+5)+(x-1) = x^2+x+4$

Write the opposite of the expression in parentheses.

$-(x^2+8) = -x^2-8 \qquad -(3x^2-5x+2) = -3x^2+5x-2$

Subtract.

$3x + {}^{+}5x = 8x \qquad (3x^2+5x)+(-x^2+{}^{-}4x) = 2x^2+x$

$^{-}2a^2 + {}^{+}6a^2 = -8a^2 \qquad (x^2+7x+8)+(-x^2+2x+5) = 9x+13$

Multiply. Make sure your answer is in simplest form.

$(x+7)(x-1) = x^2+6x-7 \qquad 2x^2(x-5) = 2x^3-10x^2$

$8x^2 \cdot \frac{1}{2x} = 4x \qquad 5y^2\left(\frac{1}{y}+\frac{2}{5}\right) = \frac{5y^2}{y} + \frac{10y^2}{5} = 5y+2y^2$

Factor completely.

$105 = 3 \cdot 5 \cdot 7$ (105 → 3·35 → 5·7)

$168 = 2 \cdot 2 \cdot 2 \cdot 3 \cdot 7$ (168 → 8·21; 8 → 2·4 → 2·2; 21 → 3·7)

$3x^2-27 = 3(x^2-9) = 3(x+3)(x-3)$

$x^2+3x-88 = (x-8)(x+11)$

Book 7, Page 2

Adding Fractions with a Common Denominator

When two fractions have the same denominator we say that they have a **common denominator**. Adding fractions with a common denominator is easy. Denominators tell *what kinds* of numbers or expressions are being added. Numerators tell *how many*, so to add fractions we just add the numerators and keep the denominators the same.

$$\frac{3}{4}+\frac{2}{4}=\frac{5}{4} \qquad \frac{4}{a}+\frac{-7}{a}=\frac{-3}{a}$$

Add.

$\frac{5}{8}+\frac{2}{8}=\frac{7}{8}$	$\frac{1}{5}+\frac{3}{5}=\frac{4}{5}$	$\frac{2}{9}+\frac{3}{9}=\frac{5}{9}$
$\frac{-7}{6}+\frac{-4}{6}=\frac{-11}{6}$	$\frac{16}{3}+\frac{-2}{3}=\frac{14}{3}$	$\frac{1}{8}+\frac{1}{8}+\frac{3}{8}=\frac{5}{8}$
$\frac{1}{x}+\frac{4}{x}=\frac{5}{x}$	$\frac{5}{a}+\frac{3}{a}=\frac{8}{a}$	$\frac{1}{5b}+\frac{3}{5b}=\frac{4}{5b}$
$\frac{4}{7y}+\frac{5}{7y}=\frac{9}{7y}$	$\frac{-6}{x^2}+\frac{1}{x^2}=\frac{-5}{x^2}$	$\frac{7}{xy}+\frac{-4}{xy}=\frac{3}{xy}$
$\frac{1}{x+3}+\frac{4}{x+3}=\frac{5}{x+3}$	$\frac{12}{a-1}+\frac{3}{a-1}=\frac{15}{a-1}$	$\frac{-7}{x^2-8}+\frac{2}{x^2-8}=\frac{-5}{x^2-8}$
$\frac{n}{3}+\frac{5}{3}=\frac{n+5}{3}$	$\frac{x}{4}+\frac{y}{4}=\frac{x+y}{4}$	$\frac{a}{5}+\frac{-3a}{5}=\frac{-2a}{5}$
$\frac{4x}{3}+\frac{3y}{3}=\frac{4x+3y}{3}$	$\frac{a^2}{9}+\frac{b^2}{9}=\frac{a^2+b^2}{9}$	$\frac{-x}{3}+\frac{8x}{3}=\frac{7x}{3}$

Book 7, Page 3

Sometimes after adding we can simplify the answer.

$$\frac{8}{9}+\frac{7}{9}=\frac{15}{9}=\frac{5}{3}$$

$$\frac{-2x}{5y}+\frac{7x}{5y}=\frac{5x}{5y}=\frac{x}{y}$$

$$\frac{x}{x^2-4}+\frac{2}{x^2-4}=\frac{x+2}{x^2-4}=\frac{x+2}{(x+2)(x-2)}=\frac{1}{x-2}$$

Add. Be sure to simplify each answer.

$\frac{3}{12}+\frac{5}{12}=\frac{8}{12}=\frac{2}{3}$	$\frac{5}{8}+\frac{-1}{8}=\frac{4}{8}=\frac{1}{2}$
$\frac{5}{9}+\frac{6}{9}+\frac{7}{9}=\frac{18}{9}=2$	$\frac{4}{3t}+\frac{2}{3t}=\frac{6}{3t}=\frac{2}{t}$
$\frac{3}{2x}+\frac{7}{2x}=\frac{10}{2x}=\frac{5}{x}$	$\frac{-5}{4x}+\frac{7}{4x}=\frac{2}{4x}=\frac{1}{2x}$
$\frac{3y}{y+4}+\frac{12}{y+4}=\frac{3y+12}{y+4}=\frac{3(y+4)}{y+4}=3$	$\frac{3x}{4x+12}+\frac{9}{4x+12}=\frac{3x+9}{4x+12}=\frac{3(x+3)}{4(x+3)}=\frac{3}{4}$

$$\frac{a}{a^2-1}+\frac{1}{a^2-1}=\frac{a+1}{a^2-1}=\frac{a+1}{(a+1)(a-1)}=\frac{1}{a-1}$$

$$\frac{x}{x^2+6x+8}+\frac{4}{x^2+6x+8}=\frac{x+4}{x^2+6x+8}=\frac{x+4}{(x+2)(x+4)}=\frac{1}{x+2}$$

$$\frac{5n}{n^2+2n}+\frac{5n}{n^2+2n}=\frac{10n}{n^2+2n}=\frac{10n}{n(n+2)}=\frac{10}{n+2}$$

$$\frac{b^2}{b^2+2b-3}+\frac{-9}{b^2+2b-3}=\frac{b^2-9}{b^2+2b-3}=\frac{(b+3)(b-3)}{(b+3)(b-1)}=\frac{b-3}{b-1}$$

Key to Algebra – ANSWERS

Book 7, Page 4

Add. Make sure each answer is in simplest form.

$$\frac{3}{y} + \frac{^-8}{y} + \frac{2}{y} = \frac{^-3}{y}$$

$$\frac{3x}{2x} + \frac{5y}{2x} = \frac{3x+5y}{2x}$$

$$\frac{x^2}{x^2+6x} + \frac{5x}{x^2+6x} + \frac{^-6}{x^2+6x} = \frac{x^2+5x-6}{x^2+6x} = \frac{(x+6)(x-1)}{x(x+6)} = \frac{x-1}{x}$$

$$\frac{x^2}{x^2-9} + \frac{3x}{x^2-9} = \frac{x^2+3x}{x^2-9} = \frac{x(x+3)}{(x+3)(x-3)} = \frac{x}{x-3}$$

$$\frac{^-5}{x+4} + \frac{^-3}{x+4} = \frac{-8}{x+4}$$

$$\frac{3x^2}{x+4} + \frac{12x}{x+4} = \frac{3x^2+12x}{x+4} = \frac{3x(x+4)}{x+4} = 3x$$

$$\frac{5x^2}{x^2-4x+3} + \frac{^-5x}{x^2-4x+3} = \frac{5x^2-5x}{x^2-4x+3} = \frac{5x(x-1)}{(x-1)(x-3)} = \frac{5x}{x-3}$$

$$\frac{11x}{3x-6} + \frac{4x}{3x-6} = \frac{15x}{3x-6} = \frac{15x}{3(x-2)} = \frac{5x}{x-2}$$

$$\frac{^-x}{x^2+4} + \frac{3x}{x^2+4} + \frac{^-2x}{x^2+4} = \frac{0}{x^2+4} = 0$$

$$\frac{2a}{a^2+2a+1} + \frac{2}{a^2+2a+1} = \frac{2a+2}{a^2+2a+1} = \frac{2(a+1)}{(a+1)(a+1)} = \frac{2}{a+1}$$

Book 7, Page 5

The numerators of these rational expressions are polynomials. Add. Then simplify the answer if you can.

add add

$$\frac{x^2+5}{x-3} + \frac{2x^2-4}{x-3} = \frac{3x^2+1}{x-3}$$

This can't be factored, so the answer can't be simplified.

$$\frac{x+10}{x+2} + \frac{x-6}{x+2} = \frac{2x+4}{x+2} = \frac{2(x+2)}{x+2} = 2$$

$$\frac{2x+3}{x+1} + \frac{x+4}{x+1} = \frac{3x+7}{x+1}$$

$$\frac{a+2b}{a+b} + \frac{2a+b}{a+b} = \frac{3a+3b}{a+b} = \frac{3(a+b)}{a+b} = 3$$

$$\frac{y^2+4}{y+1} + \frac{y^2+2y}{y+1} = \frac{2y^2+2y+4}{y+1} = \frac{2(y^2+y+2)}{y+1}$$

$$\frac{2-y^2}{3y} + \frac{y^2+7}{3y} = \frac{9}{3y} = \frac{3}{y}$$

$$\frac{2x-3}{x+6} + \frac{x-1}{x+6} = \frac{3x-4}{x+6}$$

$$\frac{x^2+7x}{3x(x+2)} + \frac{2x^2-x}{3x(x+2)} = \frac{3x^2+6x}{3x(x+2)} = \frac{3x(x+2)}{3x(x+2)} = 1$$

$$\frac{x^2+4x+3}{(x+1)(x-3)} + \frac{-x^2-x}{(x+1)(x-3)} = \frac{3x+3}{(x+1)(x-3)} = \frac{3(x+1)}{(x+1)(x-3)} = \frac{3}{x-3}$$

$$\frac{2x^2+5}{(x+1)(x-3)} + \frac{7+x-x^2}{(x+1)(x-3)} = \frac{x^2+x+12}{(x+1)(x-3)}$$

Book 7, Page 6

The Opposite of a Fraction

Subtracting rational numbers and expressions works the same way as subtracting integers. To subtract, we add the opposite — so we need to know how to find the opposite of a fraction.

The **opposite of a fraction** has the *opposite numerator* but the *same denominator* as the original fraction.

$-\frac{3}{4} = \frac{^-3}{4}$ "The opposite of $\frac{3}{4}$ is $\frac{^-3}{4}$."

$-\frac{^-6}{x} = \frac{6}{x}$ "The opposite of $\frac{^-6}{x}$ is $\frac{6}{x}$."

$-\frac{x-1}{x^2+3} = \frac{-x+1}{x^2+3}$ "The opposite of $\frac{x-1}{x^2+3}$ is $\frac{-x+1}{x^2+3}$."

Write the opposite of each fraction.

$-\frac{5}{8} = \frac{^-5}{8}$	$-\frac{x-1}{3} = \frac{-x+1}{3}$	$-\frac{a+3}{a-3} = \frac{-a-3}{a-3}$
$-\frac{^-5}{8} = \frac{5}{8}$	$-\frac{x+1}{3} = \frac{-x-1}{3}$	$-\frac{x^2-5}{x-1} = \frac{-x^2+5}{x-1}$
$-\frac{3}{x} = \frac{^-3}{x}$	$-\frac{2x+4}{x} = \frac{-2x-4}{x}$	$-\frac{x^2+2x-3}{x} = \frac{-x^2-2x+3}{x}$
$-\frac{^-3}{x} = \frac{3}{x}$	$-\frac{2x-4}{x} = \frac{-2x+4}{x}$	$-\frac{3x^2-x+4}{x^2-1} = \frac{-3x^2+x-4}{x^2-1}$
$-\frac{^-x}{6} = \frac{x}{6}$	$-\frac{4}{x-2} = \frac{-4}{x-2}$	$-\frac{5}{x^2-2x+1} = \frac{-5}{x^2-2x+1}$

Add.

$$\frac{5}{8} + \frac{^-5}{8} = \frac{0}{8} = 0 \qquad \frac{2x+7}{5} + \frac{-2x-7}{5} = \frac{0}{5} = 0$$

$$\frac{^-x}{2} + \frac{x}{2} = \frac{0}{2} = 0 \qquad \frac{x^2-4x+3}{x+1} + \frac{-x^2+4x-3}{x+1} = \frac{0}{x+1} = 0$$

Book 7, Page 7

Subtracting Fractions with a Common Denominator

To subtract one rational expression from another:

1. Replace the subtraction sign with an addition sign.
2. Replace the *second* fraction with its opposite.
3. Go ahead and add.

Subtract.

$\frac{2x}{5} + \frac{^-3x}{5} = \frac{-x}{5}$	$\frac{6}{8} + \frac{^-3}{8} = \frac{3}{8}$	$\frac{12}{13x} + \frac{^-5}{13x} = \frac{7}{13x}$
$\frac{2}{15} + \frac{^-11}{15} = \frac{-9}{15} = \frac{-3}{5}$	$\frac{x}{4} + \frac{^-5}{4} = \frac{x-5}{4}$	$\frac{2x}{9} + \frac{^-4x}{9} = \frac{-2x}{9}$
$\frac{8}{x+5} + \frac{^-5}{x+5} = \frac{3}{x+5}$	$\frac{9}{4x} + \frac{^-3x}{4x} = \frac{9-3x}{4x}$	$\frac{x}{x+y} + \frac{^-y}{x+y} = \frac{x-y}{x+y}$
$\frac{x+5}{6} + \frac{^-x}{6} = \frac{5}{6}$	$\frac{x+3}{9} + \frac{^-4}{9} = \frac{x-1}{9}$	$\frac{x+4}{x} + \frac{^-4}{x} = \frac{x}{x} = 1$

$\frac{3x+8}{2} + \frac{^-2x+^-5}{2} = \frac{x+3}{2}$	$\frac{2x+1}{x} + \frac{^-x+2}{x} = \frac{x+3}{x}$
$\frac{6x}{x+1} + \frac{^-2x+^-3}{x+1} = \frac{4x-3}{x+1}$	$\frac{3r+1}{2r} + \frac{^-3r+1}{2r} = \frac{2}{2r} = \frac{1}{r}$
$\frac{a+4}{a-2} + \frac{^-2a+6}{a-2} = \frac{-a+10}{a-2}$	$\frac{x^2-5}{x+3} + \frac{^-x+5}{x+3} = \frac{x^2-x}{x+3}$

$$\frac{x^2+3x-1}{2x} + \frac{^-x^2+5x+^-2}{2x} = \frac{8x-3}{2x}$$

Key to Algebra – ANSWERS

Book 7, Page 8

Subtract. Simplify the answer if you can.

$\frac{3x-4}{3x+3} + \frac{^-2x+5}{3x+3} = \frac{x+1}{3x+3} = \frac{x+1}{3(x+1)} = \frac{1}{3}$

$\frac{x+y}{2x} + \frac{^-x+y}{2x} = \frac{2y}{2x} = \frac{y}{x}$

$\frac{3x+2}{x} + \frac{^-4x+5}{x} = \frac{-x+7}{x}$

$\frac{3x}{4x-12} + \frac{^-9}{4x-12} = \frac{3x-9}{4x-12} = \frac{3(x-3)}{4(x-3)} = \frac{3}{4}$

$\frac{7x}{x^2+5x} + \frac{^-3x}{x^2+5x} = \frac{4x}{x^2+5x} = \frac{4x}{x(x+5)} = \frac{4}{x+5}$

$\frac{x^2+4x}{x+4} + \frac{^-4x+5}{x+4} = \frac{x^2+5}{x+4}$

$\frac{a+5}{ab} + \frac{^-a+5}{ab} = \frac{10}{ab}$

$\frac{^-x}{15} + \frac{^-4x}{15} = \frac{-5x}{15} = \frac{-x}{3}$

$\frac{6a}{5a^2+a} + \frac{^-a+1}{5a^2+a} = \frac{5a+1}{5a^2+a} = \frac{5a+1}{a(5a+1)} = \frac{1}{a}$

$\frac{x^2+3x-5}{10} + \frac{^-x^2+2x+^-10}{10} = \frac{5x-15}{10} = \frac{5(x-3)}{10} = \frac{x-3}{2}$

Book 7, Page 9

Here are some more rational expressions for you to add. This time you will have to multiply the factors in each numerator to find out if there are any like terms you can combine.

$\frac{5(x-3)}{7} + \frac{3(x+6)}{7} = \frac{5x-15}{7} + \frac{3x+18}{7} = \frac{8x+3}{7}$

$\frac{x(x-3)}{4} + \frac{3(x-2)}{4} = \frac{x^2-3x}{4} + \frac{3x-6}{4} = \frac{x^2-6}{4}$

$\frac{7(x+2)}{12x} + \frac{5(x+2)}{12x} = \frac{7x+14}{12x} + \frac{5x+2}{12x} = \frac{12x+24}{12x} = \frac{12(x+2)}{12x} = \frac{x+2}{x}$

$\frac{5(x+4)}{10x} + \frac{2(x-3)}{10x} = \frac{5x+20}{10x} + \frac{2x-6}{10x} = \frac{7x+14}{10x}$

$\frac{b(a+1)}{ab} + \frac{a(2b-3)}{ab} = \frac{ab+b}{ab} + \frac{2ab-3a}{ab} = \frac{3ab+b-3a}{ab}$

$\frac{3(5a)}{2a+2} + \frac{3(a+6)}{2a+2} = \frac{15a}{2a+2} + \frac{3a+18}{2a+2} = \frac{18a+18}{2a+2} = \frac{18(a+1)}{2(a+1)} = 9$

$\frac{x(x-1)}{x(x-3)} + \frac{6(x-4)}{x(x-3)} = \frac{x^2-x}{x(x-3)} + \frac{6x-24}{x(x-3)} = \frac{x^2+5x-24}{x(x-3)} = \frac{(x+8)(x-3)}{x(x-3)} = \frac{x+8}{x}$

$\frac{3(x+4)}{x^2-16} + \frac{3(x-4)}{x^2-16} = \frac{3x+12}{x^2-16} + \frac{3x-12}{x^2-16} = \frac{6x}{x^2-16}$

$\frac{6(a+1)}{a^2-4} + \frac{a(a-1)}{a^2-4} = \frac{6a+6}{a^2-4} + \frac{a^2-a}{a^2-4} = \frac{a^2+5a+6}{a^2-4} = \frac{(a+2)(a+3)}{(a+2)(a-2)} = \frac{a+3}{a-2}$

$\frac{(x+7)(x-2)}{(x+4)(x-3)} + \frac{(x-5)(x+2)}{(x+4)(x-3)}$ (numerators: $x^2+5x-14$, $x^2-3x-10$) $= \frac{2x^2+2x-24}{(x+4)(x-3)} = \frac{2(x^2+x-12)}{(x+4)(x-3)} = \frac{2(x+4)(x-3)}{(x+4)(x-3)} = 2$

Book 7, Page 10

These subtraction problems will be easiest if you multiply the expressions in the numerators first. Then change each problem to an equivalent addition problem.

$\frac{3(x+1)}{5} - \frac{2(x-3)}{5} = \frac{3x+3}{5} - \frac{2x-6}{5} = \frac{3x+3}{5} + \frac{^-2x+6}{5} = \frac{x+9}{5}$

$\frac{5x}{8} - \frac{4(x+1)}{8} = \frac{5x}{8} - \frac{4x+4}{8} = \frac{5x}{8} + \frac{^-4x-4}{8} = \frac{x-4}{8}$

$\frac{5(x-4)}{2x} - \frac{4(x-5)}{2x} = \frac{5x-20}{2x} - \frac{4x-20}{2x} = \frac{5x-20}{2x} + \frac{^-4x+20}{2x} = \frac{x}{2x} = \frac{1}{2}$

$\frac{2(x+4)}{x} - \frac{6(x-1)}{x} = \frac{2x+8}{x} - \frac{6x-6}{x} = \frac{2x+8}{x} + \frac{^-6x+6}{x} = \frac{^-4x+14}{x}$

$\frac{x^2}{12} - \frac{(x+1)(x-1)}{12} = \frac{x^2}{12} - \frac{x^2-1}{12} = \frac{x^2}{12} + \frac{^-x^2+1}{12} = \frac{1}{12}$

$\frac{5(x-1)}{x+2} - \frac{3(x-3)}{x+2} = \frac{5x-5}{x+2} - \frac{3x-9}{x+2} = \frac{5x-5}{x+2} + \frac{^-3x+9}{x+2} = \frac{2x+4}{x+2} = \frac{2(x+2)}{x+2} = 2$

$\frac{x(x+1)}{6} - \frac{(x+3)(x-3)}{6} = \frac{x^2+x}{6} - \frac{x^2-9}{6} = \frac{x^2+x}{6} + \frac{^-x^2+9}{6} = \frac{x+9}{6}$

$\frac{16}{x} - \frac{2(x+4)}{x} = \frac{16}{x} - \frac{2x+8}{x} = \frac{16}{x} + \frac{^-2x-8}{x} = \frac{^-2x+8}{x}$

$\frac{2a+b}{ab} - \frac{2(a-b)}{ab} = \frac{2a+b}{ab} - \frac{2a-2b}{ab} = \frac{2a+b}{ab} + \frac{^-2a+2b}{ab} = \frac{3b}{ab} = \frac{3}{a}$

$\frac{2c(c+3)}{c-5} - \frac{c(c+4)}{c-5} = \frac{2c^2+6c}{c-5} - \frac{c^2+4c}{c-5} = \frac{2c^2+6c}{c-5} + \frac{^-c^2-4c}{c-5} = \frac{c^2+2c}{c-5}$

$\frac{x^2+1}{y^2} - \frac{(x+1)(x-1)}{y^2} = \frac{x^2+1}{y^2} - \frac{x^2-1}{y^2} = \frac{x^2+1}{y^2} + \frac{^-x^2+1}{y^2} = \frac{2}{y^2}$

Book 7, Page 11

Adding and Subtracting Fractions with Different Denominators

Now we will add and subtract some fractions with different denominators.
To do this, we first have to find equivalent fractions with a common denominator.

In each problem below you first need to rewrite one of the fractions so that the two fractions have a common denominator. Then you can go ahead and combine.

$\frac{2\cdot 2x}{2\cdot 3} + \frac{x}{6} = \frac{4x}{6} + \frac{x}{6} = \frac{5x}{6}$	$\frac{2\cdot 1}{2\cdot 4} + \frac{3}{8} = \frac{2}{8} + \frac{3}{8} = \frac{5}{8}$
$\frac{7\cdot 2}{7\cdot 3} - \frac{3}{21} = \frac{14}{21} - \frac{3}{21} = \frac{11}{21}$	$\frac{4}{15} + \frac{2\cdot 3}{5\cdot 3} = \frac{4}{15} + \frac{6}{15} = \frac{10}{15} = \frac{2}{3}$
$\frac{2\cdot x}{2\cdot 8} + \frac{5x}{16} = \frac{2x}{16} + \frac{5x}{16} = \frac{7x}{16}$	$\frac{3\cdot 4}{3\cdot 5x} - \frac{4}{15x} = \frac{12}{15x} - \frac{4}{15x} = \frac{8}{15x}$
$\frac{x}{3x} + \frac{2\cdot 3}{x\cdot 3} = \frac{x}{3x} + \frac{6}{3x} = \frac{x+6}{3x}$	$\frac{5}{2x} + \frac{2\cdot 9}{2\cdot x} = \frac{5}{2x} + \frac{18}{2x} = \frac{23}{2x}$

$\frac{ab\cdot 2}{ab\cdot ab} + \frac{3}{a^2b^2} = \frac{2ab}{a^2b^2} + \frac{3}{a^2b^2} = \frac{2ab+3}{a^2b^2}$

$\frac{5x+6}{x^2} - \frac{5\cdot x}{x\cdot x} = \frac{5x+6}{x^2} + \frac{^-5x}{x^2} = \frac{6}{x^2}$

$\frac{6\cdot 2x}{6\cdot 3} + \frac{x\cdot 2}{9\cdot 2} + \frac{5x\cdot 3}{6\cdot 3} = \frac{12x}{18} + \frac{2x}{18} + \frac{15x}{18} = \frac{29x}{18}$

$\frac{2\cdot b}{a\cdot b} + \frac{3}{ab} + \frac{4\cdot a}{b\cdot a} = \frac{2b}{ab} + \frac{3}{ab} + \frac{4a}{ab} = \frac{4a+2b+3}{ab}$

$\frac{y\cdot 1}{y\cdot y} - \frac{4}{y^2} + \frac{3\cdot y}{y\cdot y} = \frac{y}{y^2} + \frac{^-4}{y^2} + \frac{3y}{y^2} = \frac{4y-4}{y^2}$

Key to Algebra – ANSWERS

Book 7, Page 12

Add or subtract.

$$\frac{3}{x+2} + \frac{x}{2(x+2)} = \frac{2\cdot3}{2(x+2)} + \frac{x}{2(x+2)} = \frac{6}{2(x+2)} + \frac{x}{2(x+2)} = \frac{6+x}{2(x+2)}$$

$$\frac{5}{x-7} + \frac{3}{4(x-7)} = \frac{4\cdot5}{4\cdot(x-7)} + \frac{3}{4(x-7)} = \frac{20}{4(x-7)} + \frac{3}{4(x-7)} = \frac{23}{4(x-7)}$$

$$\frac{5x}{(x+4)(x-4)} + \frac{1}{x-4} = \frac{5x}{(x+4)(x-4)} + \frac{(x+4)\cdot1}{(x+4)(x-4)} = \frac{5x}{(x+4)(x-4)} + \frac{x+4}{(x+4)(x-4)} = \frac{6x+4}{(x+4)(x-4)}$$

$$\frac{3}{x+5} + \frac{6}{(x+2)(x+5)} = \frac{3(x+2)}{(x+5)(x+2)} + \frac{6}{(x+2)(x+5)} = \frac{3x+6}{(x+2)(x+5)} + \frac{6}{(x+2)(x+5)} = \frac{3x+12}{(x+2)(x+5)}$$

$$\frac{6}{7} - \frac{5}{\cancel{7x-14}\;7(x-2)} = \frac{6(x-2)}{7(x-2)} - \frac{5}{7(x-2)} = \frac{6x-12}{7(x-2)} + \frac{-5}{7(x-2)} = \frac{6x-17}{7(x-2)}$$

$$\frac{(x+2)\,2}{(x+2)\,x} + \frac{5}{\cancel{x^2+2x}\;x(x+2)} = \frac{2x+4}{x(x+2)} + \frac{5}{x(x+2)} = \frac{2x+9}{x(x+2)}$$

$$\frac{3x+1}{\cancel{x^2+6x+9}\;(x+3)(x+3)} - \frac{2(x+3)}{(x+3)(x+3)} = \frac{3x+1}{(x+3)(x+3)} + \frac{-2x+-6}{(x+3)(x+3)} = \frac{x-5}{(x+3)(x+3)}$$

$$\frac{(x+6)\,10}{(x+6)(x-6)} - \frac{4}{\cancel{x^2-36}\;(x+6)(x-6)} = \frac{10x+60}{(x+6)(x-6)} + \frac{-4}{(x+6)(x-6)} = \frac{10x+56}{(x+6)(x-6)}$$

$$\frac{x}{\cancel{x^2-9}\;(x+3)(x-3)} + \frac{2(x+3)}{(x-3)(x+3)} = \frac{x}{(x+3)(x-3)} + \frac{2x+6}{(x+3)(x-3)} = \frac{3x+6}{(x+3)(x-3)}$$

$$\frac{10a}{\cancel{a^2+6a}\;a(a+6)} - \frac{3a}{(a+6)a} = \frac{10a}{a(a+6)} + \frac{-3a}{a(a+6)} = \frac{7\cancel{a}}{\cancel{a}(a+6)} = \frac{7}{a+6}$$

$$\frac{3x}{\cancel{2x^2-8x}\;2x(x-4)} + \frac{2\cdot2x}{(x-4)\cdot2x} = \frac{3x}{2x(x-4)} + \frac{4x}{2x(x-4)} = \frac{7\cancel{x}}{2\cancel{x}(x-4)} = \frac{7}{2(x-4)}$$

Book 7, Page 13

Sometimes denominators that look different are really equivalent or opposites. Be on the lookout for these. If they are opposites, it is easy to make both denominators the same. All we have to do is to multiply one fraction by $\frac{-1}{-1}$.

Add or subtract. Look for denominators that are equivalent or opposites.

$$\frac{3}{2x+5} + \frac{4x}{5+2x} = \frac{3+4x}{5+2x}$$

equivalent

$$\frac{7x}{x-3} + \frac{5x}{3-x} = \frac{7x}{x-3} + \frac{5x(-1)}{(3-x)(-1)} = \frac{7x}{x-3} + \frac{-5x}{x-3} = \frac{2x}{x-3}$$

opposites

$$\frac{4}{3x+1} + \frac{3}{1+3x} = \frac{7}{3x+1}$$

$$\frac{5(-1)}{(2-x)(-1)} + \frac{x}{x-2} = \frac{-5}{x-2} + \frac{x}{x-2} = \frac{x-5}{x-2}$$

$$\frac{2x}{3x-1} + \frac{(5x)(-1)}{(1-3x)(-1)} = \frac{2x}{3x-1} + \frac{-5x}{3x-1} = \frac{-3x}{3x-1}$$

$$\frac{7x}{x^2-1} + \frac{7(-1)}{(1-x^2)(-1)} = \frac{7x}{x^2-1} + \frac{-7}{x^2-1} = \frac{7x-7}{x^2-1} = \frac{7\cancel{(x-1)}}{(x+1)\cancel{(x-1)}} = \frac{7}{x+1}$$

$$\frac{2x+8}{x+5} + \frac{-x+-3}{5+x} = \frac{x+5}{x+5} = 1$$

$$\frac{x}{x-3} + \frac{2x(-1)}{-x+3(-1)} = \frac{x}{x-3} + \frac{-2x}{x-3} = \frac{-x}{x-3}$$

$$\frac{(12)(-1)}{(4-x)(-1)} + \frac{x+8}{x-4} = \frac{-12}{x-4} + \frac{-x+-8}{x-4} = \frac{-x-20}{x-4}$$

Book 7, Page 14

Combining Integers and Rational Expressions

In arithmetic we can write $2 + \frac{3}{4}$ as $2\frac{3}{4}$ or we can combine these into a single fraction, $\frac{11}{4}$. In algebra we can also combine an integer with a rational expression.

Combine.

$$1 + \frac{3}{x} = \frac{1\cdot x}{x} + \frac{3}{x} = \frac{x}{x} + \frac{3}{x} = \frac{x+3}{x}$$

$$1 + \frac{x}{3} = \frac{1\cdot3}{3} + \frac{x}{3} = \frac{3}{3} + \frac{x}{3} = \frac{x+3}{3}$$

$$2 - \frac{3}{4} = \frac{2\cdot4}{4} - \frac{3}{4} = \frac{8}{4} - \frac{3}{4} = \frac{5}{4}$$

$$8 - \frac{3}{4} = \frac{8\cdot4}{4} - \frac{3}{4} = \frac{32}{4} - \frac{3}{4} = \frac{29}{4}$$

$$2 + \frac{3}{x-1} = \frac{2(x-1)}{x-1} + \frac{3}{x-1} = \frac{2x-2}{x-1} + \frac{3}{x-1} = \frac{2x+1}{x-1}$$

$$5 + \frac{1}{x+2} = \frac{5(x+2)}{x+2} + \frac{1}{x+2} = \frac{5x+10}{x+2} + \frac{1}{x+2} = \frac{5x+11}{x+2}$$

$$3 + \frac{x}{x-1} = \frac{3(x-1)}{x-1} + \frac{x}{x-1} = \frac{3x-3}{x-1} + \frac{x}{x-1} = \frac{4x-3}{x-1}$$

$$\frac{2x}{x+3} - 1 = \frac{2x}{x+3} - \frac{x+3}{x+3} = \frac{2x}{x+3} + \frac{-x-3}{x+3} = \frac{x-3}{x+3}$$

$$\frac{x+4}{x+2} + 2 = \frac{x+4}{x+2} + \frac{2(x+2)}{x+2} = \frac{x+4}{x+2} + \frac{2x+4}{x+2} = \frac{3x+8}{x+2}$$

$$\frac{x}{y} + 4 = \frac{x}{y} + \frac{4y}{y} = \frac{x+4y}{y}$$

$$\frac{x}{x-y} + 1 = \frac{x}{x-y} + \frac{x-y}{x-y} = \frac{2x-y}{x-y}$$

$$\frac{x^2\cdot1}{x^2\cdot1} + \frac{1\cdot x}{x\cdot x} - \frac{1}{x^2} = \frac{x^2}{x^2} + \frac{x}{x^2} - \frac{1}{x^2} = \frac{x^2+x-1}{x^2}$$

$$1 + \frac{1}{x} + \frac{1}{x^2} = \frac{x^2}{x^2} + \frac{x}{x^2} + \frac{1}{x^2} = \frac{x^2+x+1}{x^2}$$

$$\frac{1}{x^2} - \frac{2\cdot x}{x\cdot x} + \frac{3\cdot x^2}{1\cdot x^2} = \frac{3x^2-2x+1}{x^2}$$

$$\frac{1}{4} - 1 + \frac{3}{4} = \frac{1}{4} + \frac{-4}{4} + \frac{3}{4} = \frac{0}{4} = 0$$

Book 7, Page 15

Combining Polynomials and Rational Expressions

We can combine polynomials and rational expressions, too. First we have to write the polynomial as a rational expression with the same denominator as the other expression.

Combine.

$$5y + \frac{2}{y} = \frac{y\cdot5y}{y} + \frac{2}{y} = \frac{5y^2}{y} + \frac{2}{y} = \frac{5y^2+2}{y}$$

$$2x + \frac{1}{x} = \frac{x\cdot2x}{x} + \frac{1}{x} = \frac{2x^2}{x} + \frac{1}{x} = \frac{2x^2+1}{x}$$

$$a + \frac{a}{b} = \frac{a\cdot b}{b} + \frac{a}{b} = \frac{ab+a}{b}$$

$$b + \frac{a}{b} = \frac{b\cdot b}{b} + \frac{a}{b} = \frac{b^2+a}{b}$$

$$(x+1) + \frac{3}{x} = \frac{x(x+1)}{x} + \frac{3}{x} = \frac{x^2+x}{x} + \frac{3}{x} = \frac{x^2+x+3}{x}$$

$$\frac{x}{3} + (x-5) = \frac{x}{3} + \frac{3(x-5)}{3} = \frac{x}{3} + \frac{3x-15}{3} = \frac{4x-15}{3}$$

$$(x+3) + \frac{2x}{x-1} = \frac{(x+3)(x-1)}{x-1} + \frac{2x}{x-1} = \frac{x^2+2x-3}{x-1} + \frac{2x}{x-1} = \frac{x^2+4x-3}{x-1}$$

$$(x+1) - \frac{x^2}{x+1} = \frac{(x+1)(x+1)}{x+1} - \frac{x^2}{x+1} = \frac{x^2+2x+1}{x+1} + \frac{-x^2}{x+1} = \frac{2x+1}{x+1}$$

$$(3x-4) + \frac{x^2}{2} = \frac{2(3x-4)}{2} + \frac{x^2}{2} = \frac{6x-8}{2} + \frac{x^2}{2} = \frac{x^2+6x-8}{2}$$

$$(2x-3) - \frac{1}{2x-3} = \frac{(2x-3)(2x-3)}{2x-3} - \frac{1}{2x-3} = \frac{4x^2-12x+9}{2x-3} + \frac{-1}{2x-3} = \frac{4x^2-12x+8}{2x-3}$$

Key to Algebra – ANSWERS

Book 7, Page 16

More Adding and Subtracting Fractions with Different Denominators

For each problem on this page you first need to rewrite *all* the fractions so that they have a common denominator. You can always find a common denominator by multiplying the denominators together.

$\frac{3\cdot2}{3\cdot5}+\frac{1\cdot5}{3\cdot5}=\frac{6}{15}+\frac{5}{15}=\frac{11}{15}$	$\frac{3\cdot1}{3\cdot2}+\frac{2\cdot2}{3\cdot2}=\frac{3}{6}+\frac{4}{6}=\frac{7}{6}$
$\frac{4\cdot8}{7\cdot8}-\frac{1\cdot7}{8\cdot7}=\frac{32}{56}-\frac{7}{56}=\frac{25}{56}$	$\frac{9\cdot3}{9\cdot8}-\frac{7\cdot8}{9\cdot8}=\frac{27}{72}+\frac{-56}{72}=\frac{-29}{72}$
$\frac{2x\cdot4}{5\cdot4}-\frac{x\cdot5}{4\cdot5}=\frac{8x}{20}-\frac{5x}{20}=\frac{3x}{20}$	$\frac{x\cdot7}{x\cdot y}+\frac{5\cdot y}{x\cdot y}=\frac{7x}{xy}+\frac{5y}{xy}=\frac{7x+5y}{xy}$
$\frac{14\cdot1}{14\cdot3}+\frac{5\cdot3}{14\cdot3}=\frac{14}{42}+\frac{15}{42}=\frac{29}{42}$	$\frac{b\cdot7}{b\cdot a}+\frac{3\cdot a}{b\cdot a}=\frac{7b}{ab}+\frac{3a}{ab}=\frac{7b+3a}{ab}$
$\frac{2y\cdot4}{2y\cdot x}-\frac{3\cdot x}{2y\cdot x}=\frac{8y}{2xy}-\frac{3x}{2xy}=\frac{8y-3x}{2xy}$	$\frac{5\cdot x}{5\cdot3}-\frac{4\cdot3}{5\cdot3}=\frac{5x}{15}+\frac{-12}{15}=\frac{5x-12}{15}$
$\frac{x\cdot3x}{x\cdot5}+\frac{4\cdot5}{x\cdot5}=\frac{3x^2}{5x}+\frac{20}{5x}=\frac{3x^2+20}{5x}$	$\frac{y\cdot8}{y\cdot x}-\frac{7\cdot x}{y\cdot x}=\frac{8y}{xy}+\frac{-7x}{xy}=\frac{8y-7x}{xy}$
$\frac{x\cdot10}{3\cdot10}+\frac{x\cdot15}{2\cdot15}+\frac{x\cdot6}{5\cdot6}=\frac{10x}{30}+\frac{15x}{30}+\frac{6x}{30}=\frac{31x}{30}$	$\frac{3\cdot bc}{a\cdot bc}+\frac{4\cdot ac}{b\cdot ac}+\frac{1\cdot ab}{c\cdot ab}=\frac{3bc+4ac+ab}{abc}$
$\frac{3\cdot9}{3\cdot4x}+\frac{x\cdot4x}{3\cdot4x}=\frac{27}{12x}+\frac{4x^2}{12x}=\frac{4x^2+27}{12x}$	$\frac{y^2\cdot2}{y^2\cdot x^2}-\frac{1\cdot x^2}{y^2\cdot x^2}=\frac{2y^2}{x^2y^2}+\frac{-x^2}{x^2y^2}=\frac{2y^2-x^2}{x^2y^2}$
$\frac{3x\cdot3x}{3x\cdot5y}+\frac{5y\cdot5y}{3x\cdot5y}=\frac{9x^2}{15xy}+\frac{25y^2}{15xy}=\frac{9x^2+25y^2}{15xy}$	$\frac{3\cdot4}{3\cdot a}-\frac{2\cdot a}{3\cdot a}=\frac{12}{3a}-\frac{2a}{3a}=\frac{12-2a}{3a}$
$\frac{x^2\cdot5}{x^2\cdot12}-\frac{1\cdot12}{x^2\cdot12}=\frac{5x^2}{12x^2}+\frac{-12}{12x^2}=\frac{5x^2-12}{12x^2}$	$\frac{2x\cdot8}{2x\cdot3}+\frac{7\cdot3}{2x\cdot3}=\frac{16x}{6x}+\frac{21}{6x}=\frac{16x+21}{6x}$

Book 7, Page 17

Find a common denominator. Then add or subtract.

$$\frac{3x}{7}+\frac{2}{x+5}=\frac{3x(x+5)}{7(x+5)}+\frac{7\cdot2}{7(x+5)}=\frac{3x^2+15x}{7(x+5)}+\frac{14}{7(x+5)}=\frac{3x^2+15x+14}{7(x+5)}$$

$$\frac{3}{8}-\frac{5}{a-2}=\frac{3(a-2)}{8(a-2)}-\frac{8\cdot5}{8(a-2)}=\frac{3a-6}{8(a-2)}+\frac{-40}{8(a-2)}=\frac{3a-46}{8(a-2)}$$

$$\frac{5}{x+2}+\frac{3}{4x}=\frac{5\cdot4x}{(x+2)4x}+\frac{3(x+2)}{4x(x+2)}=\frac{20x}{4x(x+2)}+\frac{3x+6}{4x(x+2)}=\frac{23x+6}{4x(x+2)}$$

$$\frac{2}{x+1}+\frac{2}{5x}=\frac{5x(2)}{5x(x+1)}+\frac{2(x+1)}{5x(x+1)}=\frac{10x}{5x(x+1)}+\frac{2x+2}{5x(x+1)}=\frac{12x+2}{5x(x+1)}$$

$$\frac{x-1}{x+3}+\frac{2}{x+1}=\frac{(x-1)(x+1)}{(x+3)(x+1)}+\frac{(x+3)\cdot2}{(x+3)(x+1)}=\frac{x^2-1}{(x+3)(x+1)}+\frac{2x+6}{(x+3)(x+1)}=\frac{x^2+2x+5}{(x+3)(x+1)}$$

$$\frac{x+4}{x}-\frac{x}{x+3}=\frac{(x+4)(x+3)}{x(x+3)}-\frac{x\cdot x}{x(x+3)}=\frac{x^2+7x+12}{x(x+3)}+\frac{-x^2}{x(x+3)}=\frac{7x+12}{x(x+3)}$$

$$\frac{x+3}{x+5}+\frac{x}{x+2}=\frac{(x+2)(x+3)}{(x+2)(x+5)}+\frac{x(x+5)}{(x+2)(x+5)}=\frac{x^2+5x+6}{(x+2)(x+5)}+\frac{x^2+5x}{(x+2)(x+5)}=\frac{2x^2+10x+6}{(x+2)(x+5)}$$

$$\frac{x+4}{x-4}+\frac{3}{x+3}=\frac{(x+3)(x+4)}{(x+3)(x-4)}+\frac{3(x-4)}{(x+3)(x-4)}=\frac{x^2+7x+12}{(x+3)(x-4)}+\frac{3x-12}{(x+3)(x-4)}=\frac{x^2+10x}{(x+3)(x-4)}$$

$$\frac{x-4}{x+6}-\frac{x}{x+7}=\frac{(x+7)(x-4)}{(x+7)(x+6)}-\frac{x(x+6)}{(x+7)(x+6)}=\frac{x^2+3x-28}{(x+7)(x+6)}+\frac{-x^2-6x}{(x+7)(x+6)}=\frac{-3x-28}{(x+7)(x+6)}$$

$$\frac{2x}{2x+1}+\frac{x-1}{x+4}=\frac{2x(x+4)}{(2x+1)(x+4)}+\frac{(x-1)(2x+1)}{(x+4)(2x+1)}=\frac{2x^2+8x}{(2x+1)(x+4)}+\frac{2x^2-x-1}{(x+4)(2x+1)}=\frac{4x^2+7x-1}{(2x+1)(x+4)}$$

$$\frac{x-5}{x+1}-\frac{x+3}{x-1}=\frac{(x-1)(x-5)}{(x-1)(x+1)}-\frac{(x+3)(x+1)}{(x-1)(x+1)}=\frac{x^2-6x+5}{(x-1)(x+1)}+\frac{-x^2-4x-3}{(x-1)(x+1)}=\frac{-10x+2}{(x-1)(x+1)}$$

Book 7, Page 18

Least Common Denominators

The product of the two denominators is not always the least or simplest denominator we could use. Think about adding $\frac{5}{12}$ and $\frac{7}{8}$.

If we use 12 times 8, or 96, as a common denominator, we get

$$\frac{5}{12}+\frac{7}{8}=\frac{5\cdot8}{12\cdot8}+\frac{12\cdot7}{12\cdot8}=\frac{40}{96}+\frac{84}{96}=\frac{124}{96}$$

We could have used 24 as a common denominator.

$$\frac{5}{12}+\frac{7}{8}=\frac{5\cdot2}{12\cdot2}+\frac{3\cdot7}{3\cdot8}=\frac{10}{24}+\frac{21}{24}=\frac{31}{24}$$

The answers are equivalent, but $\frac{31}{24}$ is simpler.

Add or subtract each pair of fractions. Try to find the **least common denominator**.

$$\frac{5\cdot11}{5\cdot6}-\frac{4\cdot2}{15\cdot2}=\frac{55}{30}+\frac{-8}{30}=\frac{47}{30}$$

$$\frac{9\cdot3}{9\cdot20}+\frac{8\cdot4}{45\cdot4}=\frac{27}{180}+\frac{32}{180}=\frac{59}{180}$$

$$\frac{6\cdot a}{6\cdot20}-\frac{b\cdot5}{24\cdot5}=\frac{6a}{120}-\frac{5b}{120}=\frac{6a-5b}{120}$$

$$\frac{2\cdot1}{2\cdot12x}+\frac{5\cdot3}{8x\cdot3}=\frac{2}{24x}+\frac{15}{24x}=\frac{17}{24x}$$

$$\frac{x\cdot3}{x\cdot2x}+\frac{2\cdot2}{x^2\cdot2}=\frac{3x}{2x^2}+\frac{4}{2x^2}=\frac{3x+4}{2x^2}$$

$$\frac{3y\cdot5}{3y\cdot2x^2}+\frac{3\cdot x}{6xy\cdot x}=\frac{15y}{6x^2y}+\frac{3x}{6x^2y}=\frac{15y+3x}{6x^2y}$$

$$\frac{y\cdot1}{y\cdot x^2}+\frac{4\cdot x}{xy\cdot x}=\frac{y+4x}{x^2y}$$

$$\frac{5\cdot5}{5\cdot6x}-\frac{1\cdot3x}{10\cdot3x}=\frac{25}{30x}+\frac{-3x}{30x}=\frac{25-3x}{30x}$$

Book 7, Page 19

Did you discover a way to find the least common denominator? Here is a method that always works. We factor the denominators. Then we multiply each denominator by the factors of the other denominator which are missing. We multiply the numerator by those factors, too.

Find the least common denominator. Then combine.

$$\frac{5}{12}+\frac{7}{8}=\frac{5}{2\cdot2\cdot3}+\frac{7}{2\cdot2\cdot2}=\frac{5\cdot2}{2\cdot2\cdot3\cdot2}+\frac{7\cdot3}{2\cdot2\cdot2\cdot3}=\frac{10}{24}+\frac{21}{24}=\frac{31}{24}$$

(The other denominator has 3 2's. This one needs another 2.) (The other denominator has a 3, so this one needs a 3.)

$$\frac{8}{35}+\frac{2}{45}=\frac{8}{5\cdot7}+\frac{2}{3\cdot3\cdot5}=\frac{8\cdot3\cdot3}{5\cdot7\cdot3\cdot3}+\frac{2\cdot7}{3\cdot3\cdot5\cdot7}=\frac{72}{315}+\frac{14}{315}=\frac{86}{315}$$

$$\frac{2}{9}-\frac{2}{15}=\frac{2}{3\cdot3}-\frac{2}{3\cdot5}=\frac{2\cdot5}{3\cdot3\cdot5}-\frac{2\cdot3}{3\cdot5\cdot3}=\frac{10}{45}-\frac{6}{45}=\frac{4}{45}$$

$$\frac{3}{ab}+\frac{4}{b^2c}=\frac{3}{ab}+\frac{4}{bbc}=\frac{3\cdot bc}{ab\cdot bc}+\frac{4\cdot a}{a\cdot bbc}=\frac{3bc}{ab^2c}+\frac{4a}{ab^2c}=\frac{3bc+4a}{ab^2c}$$

(Needs a b and a c.) (Needs an a.)

$$\frac{2}{9x^2}+\frac{4}{15x}=\frac{2\cdot5}{3\cdot3\cdot x\cdot x\cdot5}+\frac{4\cdot3\cdot x}{3\cdot5\cdot x\cdot3\cdot x}=\frac{10}{45x^2}+\frac{12x}{45x^2}=\frac{10+12x}{45x^2}$$

$$\frac{8\cdot d}{abc\cdot d}-\frac{2\cdot a}{bcd\cdot a}=\frac{8d-2a}{abcd}$$

$$\frac{1}{14x^2}+\frac{1}{21xy}=\frac{1\cdot3\cdot y}{2\cdot7\cdot x\cdot x\cdot3\cdot y}+\frac{1\cdot2\cdot x}{3\cdot7\cdot x\cdot y\cdot2\cdot x}=\frac{3y}{42x^2y}+\frac{2x}{42x^2y}=\frac{3y+2x}{42x^2y}$$

$$\frac{3}{x}+\frac{2}{x^2}-\frac{1}{x^3}=\frac{3\cdot x\cdot x}{x\cdot x\cdot x}+\frac{2\cdot x}{x\cdot x\cdot x}+\frac{-1}{x\cdot x\cdot x}=\frac{3x^2+2x-1}{x^3}$$

$$\frac{5}{a^2b}-\frac{3}{ab^2}=\frac{5\cdot b}{a\cdot a\cdot b\cdot b}-\frac{3\cdot a}{a\cdot b\cdot b\cdot a}=\frac{5b}{a^2b^2}+\frac{-3a}{a^2b^2}=\frac{5b-3a}{a^2b^2}$$

Key to Algebra – ANSWERS

Book 7, Page 20

Find the least common denominator. Then add or subtract.

$$\frac{x}{4x+6} - \frac{2}{6x+9} = \frac{3\cdot x}{3\cdot 2(2x+3)} - \frac{2\cdot 2}{2\cdot 3(2x+3)} = \frac{3x}{6(2x+3)} + \frac{-4}{6(2x+3)} = \frac{3x-4}{6(2x+3)}$$
$2(2x+3)$, $3(2x+3)$

$$\frac{1\cdot 4}{x^2-3x} + \frac{5\cdot x}{4x-12} = \frac{4}{4x(x-3)} + \frac{5x}{4x(x-3)} = \frac{5x+4}{4x(x-3)}$$
$x(x-3)\cdot 4$, $4(x-3)\cdot x$

$$\frac{(x+1)\,3}{(x+1)\,x-1} + \frac{4}{x^2-1} = \frac{3x+3}{(x+1)(x-1)} + \frac{4}{(x+1)(x-1)} = \frac{3x+7}{(x+1)(x-1)}$$
$(x+1)(x-1)$

$$\frac{1}{3x-6} - \frac{x}{x^2-4} = \frac{1(x+2)}{3(x+2)(x-2)} + \frac{-3x}{(x+2)(x-2)\cdot 3} = \frac{-2x+2}{3(x+2)(x-2)}$$
$3(x-2)$, $(x+2)(x-2)$

$$\frac{3}{x^2-25} + \frac{5}{2x+10} = \frac{6}{(x+5)(x-5)\cdot 2} + \frac{5x-25}{2(x+5)(x-5)} = \frac{5x-19}{2(x+5)(x-5)}$$
$(x+5)(x-5)$, $2(x+5)$

$$\frac{5}{x^2-9} + \frac{2}{x^2-6x+9} = \frac{5x-15}{(x+3)(x-3)(x-3)} + \frac{2x+6}{(x-3)(x-3)(x+3)} = \frac{7x-9}{(x-3)(x-3)(x+3)}$$
$(x+3)(x-3)$, $(x-3)(x-3)$

$$\frac{4}{n^2-4} + \frac{2}{n^2-5n+6} = \frac{4n-12}{(n+2)(n-2)(n-3)} + \frac{2n+4}{(n-2)(n-3)(n+2)} = \frac{6n-8}{(n-2)(n-3)(n+2)}$$
$(n+2)(n-2)$, $(n-2)(n-3)$

$$\frac{3}{x^2+6x+5} + \frac{1}{x^2+4x+3} = \frac{3x+9}{(x+5)(x+1)(x+3)} + \frac{x+5}{(x+1)(x+3)(x+5)} = \frac{4x+14}{(x+1)(x+3)(x+5)}$$
$(x+5)(x+1)$, $(x+1)(x+3)$

$$\frac{5}{3x^2-12} + \frac{2}{4x+8} = \frac{20}{3(x+2)(x-2)4} + \frac{6x-12}{4(x+2)\cdot 3(x-2)} = \frac{2(3x+4)}{12(x+2)(x-2)} = \frac{3x+4}{6(x+2)(x-2)}$$
$3(x+2)(x-2)$, $4(x+2)$

Book 7, Page 21

Add.

$$\frac{4}{x-6} + \frac{3(-1)}{(6-x)(-1)} = \frac{4}{x-6} + \frac{-3}{x-6} = \frac{1}{x-6}$$

$$\frac{7(a+5)}{(a+2)(a+5)} + \frac{4(a+2)}{(a+5)(a+2)} = \frac{7a+35}{(a+2)(a+5)} + \frac{4a+8}{(a+2)(a+5)} = \frac{11a+43}{(a+2)(a+5)}$$

$$\frac{5\cdot x}{5(x+3)} + \frac{2}{5x+15} = \frac{5x+2}{5(x+3)}$$
$5(x+3)$

$$\frac{10\cdot y}{x^2y\cdot y} + \frac{4\cdot x}{xy^2\cdot x} = \frac{10y+4x}{x^2y^2}$$

$$\frac{2\cdot 1}{x^2-3x} + \frac{3\cdot x}{2x-6} = \frac{2+3x}{2x(x-3)}$$
$2\cdot x(x-3)$, $2(x-3)\cdot x$

$$\frac{x}{y} + \frac{4\cdot y}{1\cdot y} = \frac{x+4y}{y}$$

$$\frac{3\cdot b}{a\cdot b} + \frac{2\cdot a}{b\cdot a} + \frac{4}{ab} = \frac{3b+2a+4}{ab}$$

$$\frac{x}{x^2-1} + \frac{2}{x^2+6x-7} = \frac{x^2+7x}{(x+1)(x-1)(x+7)} + \frac{2x+2}{(x+7)(x-1)(x+1)} = \frac{x^2+9x+2}{(x+7)(x-1)(x+1)}$$
$(x+1)(x-1)$, $(x+7)(x-1)$

$$\frac{y}{3y+1} + \frac{2}{1+3y} = \frac{y+2}{3y+1}$$
$3y+1$

Book 7, Page 22

Find a common denominator and add. Then simplify.

$$\frac{3(x-2)}{3\cdot 4} + \frac{x+2}{12} = \frac{3x-6}{12} + \frac{x+2}{12} = \frac{4x-4}{12} = \frac{4(x-1)}{12} = \frac{x-1}{3}$$

$$\frac{2x+1}{20} + \frac{x+3}{15} = \frac{3(2x+1)}{3\cdot 2\cdot 2\cdot 5} + \frac{2\cdot 2(x+3)}{2\cdot 2\cdot 3\cdot 5} = \frac{10x+15}{60} = \frac{2x+3}{12}$$
$2\cdot 2\cdot 5$, $3\cdot 5$

$$\frac{2(x+6)}{2(x+5)} + \frac{x+3}{2x+10} = \frac{3x+15}{2(x+5)} = \frac{3(x+5)}{2(x+5)} = \frac{3}{2}$$

$$\frac{3\cdot 3}{3(x-2)} + \frac{x-11}{3x-6} = \frac{x-2}{3(x-2)} = \frac{1}{3}$$

$$\frac{2(a+b)}{2\cdot ab} + \frac{a-2b}{2ab} = \frac{3a}{2ab} = \frac{3}{2b}$$

$$\frac{x(x+2)}{x(x-1)} + \frac{x-4}{x^2-x} = \frac{x^2+3x-4}{x(x-1)} = \frac{(x+4)(x-1)}{x(x-1)} = \frac{x+4}{x}$$

$$\frac{x}{2x-5} + \frac{(5-x)(-1)}{(5-2x)(-1)} = \frac{x}{2x-5} + \frac{x+5}{2x-5} = \frac{2x-5}{2x-5} = 1$$

$$\frac{8}{4x+4} + \frac{x+3}{x^2+x} = \frac{8\cdot x}{4(x+1)\cdot x} + \frac{(x+3)\cdot 4}{x(x+1)\cdot 4} = \frac{8x}{4x(x+1)} + \frac{4x+12}{4x(x+1)} = \frac{12x+12}{4x(x+1)} = \frac{12(x+1)}{4x(x+1)} = \frac{3}{x}$$

$$\frac{x-6}{x^2-5x} + \frac{1}{5x-25} = \frac{5(x-6)}{5\cdot x(x-5)} + \frac{1\cdot x}{5(x-5)\cdot x} = \frac{6(x-5)}{5x(x-5)} = \frac{6}{5x}$$
$x(x-5)$, $5(x-5)$

Book 7, Page 23

Using Common Denominators to Solve Equations

Common denominators are useful for solving equations which contain rational expressions. In the equation below there are three fractions. The least common denominator for these fractions is $6x^2$. If we multiply both sides of the equation by the common denominator, we will get an equivalent equation without fractions.

$$\frac{1}{2x} + \frac{2}{x^2} = \frac{7}{6x}$$ (The least common denominator is $6x^2$.)

$$6x^2\left(\frac{1}{2x} + \frac{2}{x^2}\right) = \frac{7}{6x}\cdot 6x^2$$

$$\frac{6x^2}{1}\cdot\frac{1}{2x} + \frac{6x^2}{1}\cdot\frac{2}{x^2} = \frac{7}{6x}\cdot\frac{6x^2}{1}$$

$3x + 12 = 7x$ (3 works: $\frac{1}{6} + \frac{2}{9} = \frac{7}{18}$ is true, so it's a solution.)

$12 = 4x$

$x = 3$

Solve each equation by multiplying both sides by the least common denominator.

$$\frac{3x}{1}\left(\frac{1}{3} + \frac{2}{x}\right) = \frac{2}{3}\cdot\frac{3x}{1}$$
$$\frac{3x}{3} + \frac{3x\cdot 2}{x} = \frac{2\cdot 3x}{3}$$
$x + 6 = 2x$

$6 = x$

$x = 6$

$$\frac{10}{1}\left(\frac{x}{5} + \frac{x}{2}\right) = \frac{14}{5}\cdot\frac{10}{1}$$
$$\frac{10x}{5} + \frac{10x}{2} = \frac{14\cdot 10}{5}$$
$2x + 5x = 28$

$7x = 28$

$x = 4$

$$\frac{3}{1}\cdot\left(\frac{x}{3} + \frac{2x}{3}\right) = 4\cdot 3$$
$$\frac{3x}{3} + \frac{3\cdot 2x}{3} = 12$$
$x + 2x = 12$

$3x = 12$

$x = 4$

$$12\cdot\left(\frac{3a}{4} + \frac{a}{6}\right) = 0\cdot 12$$
$$\frac{12\cdot 3a}{4} + \frac{12a}{6} = 0$$
$9a + 2a = 0$

$11a = 0$

$a = 0$

Key to Algebra – ANSWERS

Book 7, Page 24

Solve each equation.

$8\cdot\left(\frac{x+1}{8}\right)=\left(\frac{x}{2}-\frac{1}{4}\right)\cdot 8$
$\frac{8(x+1)}{8}=\frac{^{4}8x}{2}-\frac{8^{2}}{4}$
$x+1=4x-2$
$1=3x-2$
$3=3x$
$x=1$

$24\left(\frac{x+2}{8}-\frac{x}{12}\right)=\frac{1}{2}\cdot 24$
$\frac{^{3}24(x+2)}{8}-\frac{^{2}24x}{12}=\frac{24^{12}}{2}$
$3(x+2)-2x=12$
$3x+6-2x=12$
$x+6=12$
$x=6$

$4x\cdot\left(\frac{3}{2x}-\frac{1}{x}\right)=\frac{1}{4}\cdot 4x$
$\frac{^{2}4x\cdot 3}{2x}-\frac{4x}{x}=\frac{4x}{4}$
$6-4=x$
$2=x$
$x=2$

$x\left(x+\frac{24}{x}\right)=11\cdot x$
$x^2+\frac{24x}{x}=11x$
$x^2+24=11x$
$x^2-11x+24=0$
$(x-8)(x-3)=0$
$x-8=0$ or $x-3=0$
$x=8$ or $x=3$

$2x\cdot\left(5+\frac{3}{x}\right)=\frac{7}{2}\cdot 2x$
$2x\cdot 5+\frac{2x\cdot 3}{x}=\frac{7\cdot 2x}{2}$
$10x+6=7x$
$6=-3x$
$x=-2$

$2x\cdot\left(\frac{x}{2}-\frac{7}{2}\right)=\frac{9}{x}\cdot 2x$
$\frac{2x\cdot x}{2}-\frac{2x\cdot 7}{2}=\frac{9\cdot 2x}{x}$
$x^2-7x=18$
$x^2-7x-18=0$
$(x-9)(x+2)=0$
$x-9=0$ or $x+2=0$
$x=9$ or $x=-2$

$2(x-1)\frac{3}{x-1}=\left(\frac{1}{2}-\frac{4}{x-1}\right)2(x-1)$
$\frac{2\cdot 3(x-1)}{x-1}=\frac{2(x-1)}{2}-\frac{4\cdot 2(x-1)}{x-1}$
$6=x-1-8$
$6=x-9$
$15=x$
$x=15$

$3(x+4)\left(\frac{x}{3}+\frac{10}{3(x+4)}\right)=\left(\frac{5}{x+4}\right)3(x+4)$
$\frac{3x(x+4)}{3}+\frac{3\cdot 10(x+4)}{3(x+4)}=\frac{5\cdot 3(x+4)}{x+4}$
$x(x+4)+10=15$
$x^2+4x+10=15$
$x^2+4x-5=0$
$(x+5)(x-1)=0$
$x+5=0$ or $x-1=0$
$x=-5$ or $x=1$

Book 7, Page 25

Here are some more equations with rational expressions for you to solve.

$12\left(\frac{3}{4}x+\frac{1}{3}x\right)=(x+1)12$
$\frac{^{3}12\cdot 3}{4}x+\frac{^{4}12\cdot 1}{3}x=12x+12$
$9x+4x=12x+12$
$13x=12x+12$
$x=12$

$10\left(\frac{2}{5}x-\frac{1}{10}x\right)=1\cdot 10$
$\frac{^{2}10\cdot 2}{5}x-\frac{10}{10}x=10$
$4x-x=10$
$\frac{3x}{3}=\frac{10}{3}$
$x=\frac{10}{3}$

$0.8x=2+0.7x$
(This says $\frac{8}{10}x=2+\frac{7}{10}x$, so multiply both sides by 10.)
$10(0.8x)=(2+0.7x)10$
$8x=20+7x$
$x=20$

$10\cdot(0.5+0.1x)=(0.8x-0.9)\cdot 10$
$5+1x=8x-9$
$5-7x=-9$
$-7x=-14$
$x=2$

$100\cdot(0.23x+2.25)=(0.45x-1.05)\cdot 100$
$23x+225=45x-105$
$225=22x-105$
$\frac{330}{22}=\frac{22x}{22}$
$x=15$

$100\cdot(0.7x-0.66x)=2\cdot 100$
$70x-66x=200$
$4x=200$
$x=50$

$100\cdot(0.2x-0.08x)=1.2\cdot 100$
$20x-8x=120$
$12x=120$
$x=10$

$100\cdot 1.24x=(3.2+0.6x)\cdot 100$
$124x=320+60x$
$\frac{64x}{64}=\frac{320}{64}$
$x=5$

Book 7, Page 26

Proportions

A **proportion** is a very simple rational equation which says that one fraction equals another. Each of these is a proportion:

$\frac{3}{4}=\frac{15}{20}$ $\frac{x}{2}=\frac{5}{9}$ $\frac{x+1}{x}=\frac{4}{x+2}$

Look at what happens to a proportion if we get a common denominator by multiplying the two denominators together and then use it to simplify the equation.

$\frac{3}{4}=\frac{15}{20}$ $\frac{x}{2}=\frac{5}{9}$

$\overset{20}{80}\cdot\frac{3}{4}=\frac{15}{20}\cdot\overset{4}{80}$ $\overset{9}{18}\cdot\frac{x}{2}=\frac{5}{9}\cdot\overset{2}{18}$

$20\cdot 3=15\cdot 4$ $9x=5\cdot 2$

It looks as if each denominator has "moved" to the other side of the equation.

$\frac{3}{4}\times\frac{15}{20}$ $\frac{x}{2}\times\frac{5}{9}$

$20\cdot 3=15\cdot 4$ $9x=5\cdot 2$

The arrows show where each denominator ended up. Since this is always the pattern, we usually skip the first step and go right to multiplying each numerator by the opposite denominator. We call this **cross multiplying**.

Cross multiplying makes it easy to solve proportions. Use cross multiplying to solve each proportion below.

$\frac{12}{16}=\frac{x}{20}$
$20\cdot 12=16\cdot x$
$240=16x$
$x=15$

$\frac{-10}{8}=\frac{x}{12}$
$-10\cdot 12=8x$
$\frac{-120}{8}=\frac{8x}{8}$
$x=-15$

$\frac{20}{8}=\frac{15}{x}$
$20x=8\cdot 15$
$20x=120$
$x=6$

$\frac{x}{4}=\frac{3}{8}$
$\frac{8x}{8}=\frac{12}{8}$
$x=\frac{12^{3}}{8_{2}}$
$x=\frac{3}{2}$

$\frac{4}{x}=\frac{7}{9}$
$\frac{7x}{7}=\frac{36}{7}$
$x=\frac{36}{7}$

$\frac{8}{3}=\frac{x+1}{6}$
$3(x+1)=48$
$3x+3=48$
$3x=45$
$x=15$

Book 7, Page 27

Solve each proportion.

$\frac{10}{x}=\frac{15}{x-6}$
$10(x-6)=15\cdot x$
$10x-60=15x$
$-60=5x$
$x=-12$

$\frac{x}{6}=\frac{x+8}{18}$
$18x=6(x+8)$
$18x=6x+48$
$12x=48$
$x=4$

$\frac{8}{x}=\frac{14}{x+3}$
$8(x+3)=14x$
$8x+24=14x$
$24=6x$
$x=4$

$\frac{3}{7a}=\frac{2}{5}$
$7a\cdot 2=15$
$\frac{14a}{14}=\frac{15}{14}$
$a=\frac{15}{14}$

$\frac{p}{p+4}=\frac{7}{8}$
$8p=7(p+4)$
$8p=7p+28$
$p=28$

$\frac{3}{n+2}=\frac{-3}{n-2}$
$3(n-2)=-3(n+2)$
$3n-6=-3n-6$
$6n-6=-6$
$6n=0$
$n=0$

$\frac{4}{5}=\frac{y-3}{y+3}$
$4(y+3)=5(y-3)$
$4y+12=5y-15$
$12=y-15$
$27=y$
$y=27$

$\frac{x+3}{x-1}=\frac{x-5}{x-7}$
$(x+3)(x-7)=(x-1)(x-5)$
$x^2-4x-21=x^2-6x+5$
$-4x-21=-6x+5$
$2x-21=5$
$2x=26$
$x=13$

$\frac{x}{x+6}=\frac{x-2}{x+2}$
$x(x+2)=(x+6)(x-2)$
$x^2+2x=x^2+4x-12$
$2x=4x-12$
$-2x=-12$
$x=6$

$\frac{x}{4}=\frac{1}{x}$
$x^2=4$
$x^2-4=0$
$(x-2)(x+2)=0$
$x-2=0$ | $x+2=0$
$x=2$ | $x=-2$

$\frac{x+1}{3}=\frac{2}{x}$
$x(x+1)=6$
$x^2+x=6$
$x^2+x-6=0$
$(x+3)(x-2)=0$
$x+3=0$ | $x-2=0$
$x=-3$ | $x=2$

$\frac{x}{4}=\frac{9}{x}$
$x^2=36$
$x^2-36=0$
$(x+6)(x-6)=0$
$x+6=0$ | $x-6=0$
$x=-6$ | $x=6$

$\frac{x+6}{-3}=\frac{4}{x-1}$
$(x+6)(x-1)=-12$
$x^2+5x-6=-12$
$x^2+5x+6=0$
$(x+2)(x+3)=0$
$x+2=0$ | $x+3=0$
$x=-2$ | $x=-3$

$\frac{x+7}{3}=\frac{5}{x-7}$
$(x+7)(x-7)=15$
$x^2-49=15$
$x^2-64=0$
$(x+8)(x-8)=0$
$x+8=0$ | $x-8=0$
$x=-8$ | $x=8$

$\frac{20}{x}=\frac{x}{5}$
$x^2=100$
$x^2-100=0$
$(x-10)(x+10)=0$
$x-10=0$ | $x+10=0$
$x=10$ | $x=-10$

Key to Algebra – ANSWERS

Book 7, Page 28

Ratio Problems

We can use proportions to solve many problems that involve ratios. A **ratio** is a fraction which compares one number to another. The ratio of 2 to 3 is $\frac{2}{3}$. We can write a proportion if we know that two ratios are equivalent.

Write a proportion for each problem. Then solve the proportion to find the answer.

If it takes 5 gallons of gas to drive 90 miles, how many gallons would it take to drive 144 miles?

(The ratio of gallons to miles is always the same.)

Equation: $\frac{5}{90} = \frac{x}{144}$

$144 \cdot 5 = 90 \cdot x$

$720 = 90x$

$x = 8$

Answer: It takes 8 gallons to drive 144 miles.

The cafeteria used 28 bottles of ketchup in 18 days. How much ketchup should be ordered for 45 days?

Equation: $\frac{28}{18} = \frac{x}{45}$

$28 \cdot 45 = 18 \cdot x$

$1260 = 18x$

$x = 70$

Answer: 70 bottles of ketchup should be ordered.

A ball player gets 11 hits in 40 times at bat. How many hits would you predict in 1000 times at bat?

Equation: $\frac{11}{40} = \frac{x}{1000}$

$40x = 11 \cdot 1000$

$\frac{40x}{40} = \frac{11{,}000}{40}$

$x = 275$

Answer: 275 hits

The ratio of sand to cement in concrete is 5 to 2. How many shovels of sand should be mixed with 8 shovels of cement?

Equation: $\frac{5}{2} = \frac{x}{8}$

$2x = 5 \cdot 8$

$2x = 40$

$x = 20$

Answer: 20 shovels of sand

A typist can type 300 words in 4 minutes. At that rate, how long would it take to type a 13,500 word document?

Equation: $\frac{300}{4} = \frac{13{,}500}{x}$

$300x = 13{,}500 \cdot 4$

$\frac{300x}{300} = \frac{54{,}000}{300}$

$x = 180$

Answer: 180 minutes or 3 hours

If 12 lb. of coleslaw will feed 100 people, how much would you need to feed 40 people?

Equation: $\frac{12}{100} = \frac{x}{40}$

$100x = 12 \cdot 40$

$\frac{100x}{100} = \frac{480}{100}$

$x = 4.8$

Answer: 4.8 or $4\frac{8}{10}$ lb.

Book 7, Page 29

Solve each problem by using a proportion.

The ratio of red to brown candies in a well-known brand is 2 to 3. Chris likes the red ones. Jamie prefers brown. How many will Chris be likely to get if Jamie gets 45?

Equation: C - r, J - b

$\frac{2}{3} = \frac{x}{45}$

$3x = 2 \cdot 45$

$\frac{3x}{3} = \frac{90}{3}$

$x = 30$

Answer: Chris will likely get 30.

The label on a 48-lb. bag of lawn fertilizer says it covers 15,000 square feet. How many pounds will be needed to cover 50,000 square feet?

Equation: $\frac{48}{15{,}000} = \frac{x}{50{,}000}$

$15{,}000x = 48 \cdot 50{,}000$

$\frac{15{,}000x}{15{,}000} = \frac{2{,}400{,}000}{15{,}000}$

$x = 160$

Answer: 160 pounds

A car's trip meter registers 10.2 miles for a 10-mile measured course. When the meter registers 153 miles, what distance has the car actually traveled?

Equation: $\frac{10.2}{10} = \frac{153}{x}$

$10.2x = 10 \cdot 153$

$10.2x = 1530$

$\frac{102x}{102} = \frac{15300}{102}$

$x = 150$

Answer: 150 miles

A cake recipe calls for 2 eggs to $1\frac{1}{2}$ cups of flour. How many eggs should be used with 9 cups of flour?

Equation: $\frac{2}{1\frac{1}{2}} = \frac{x}{9}$

$1\frac{1}{2}x = 2 \cdot 9$

$2 \cdot \frac{3}{2}x = 18 \cdot 2$

$3x = 36$

$x = 12$

Answer: 12 eggs

The ratio of inches to centimeters in a measurement is about 2 to 5. Maria's waist measures 70 centimeters. About what would that be in inches?

Equation: $\frac{2}{5} = \frac{x}{70}$

$5x = 2 \cdot 70$

$\frac{5x}{5} = \frac{140}{5}$

$x = 28$

Answer: about 28 inches

The ratio of left-handed people to all people is about 1 to 10. How many left-handed students would you expect to find in a high school with 450 students?

Equation: $\frac{1}{10} = \frac{x}{450}$

$10x = 1 \cdot 450$

$\frac{10x}{10} = \frac{450}{10}$

$x = 45$

Answer: 45 left-handed students

Book 7, Page 30

Percent Problems

The word **percent** means "hundredths" or "out of a hundred." A percent is a ratio with a denominator of 100 and so can be written as either a fraction or a decimal.

73%, $\frac{73}{100}$ and .73 all stand for the ratio of 73 to 100.

Often the easiest way to solve a percent problem is to write a proportion.

Write a proportion for each problem. Then solve the proportion to get the answer.

32 out of 80 people in line got tickets for the concert. What percent of the people in line were successful?

(32 out of 80 is equivalent to what number out of 100?)

Equation: $\frac{32}{80} = \frac{x}{100}$

$100 \cdot 32 = 80 \cdot x$

$3200 = 80x$

$x = 40$

Answer: 40% of the people were successful in getting tickets.

Concert tickets cost $26 each. 22% went to charity. How much went to charity?

Equation: $\frac{x}{26} = \frac{22}{100}$

$100 \cdot x = 26 \cdot 22$

$100x = 572$

$x = 5.72$

Answer: $5.72 from each ticket went to charity.

$15 of Dan's $125 wages was withheld for taxes. What percent of his wages was withheld?

Equation: $\frac{15}{125} = \frac{x}{100}$

$125x = 100 \cdot 15$

$125x = 1500$

$x = 12$

Answer: 12% of Dan's wages was withheld.

If 12% is always withheld from Dan's wages, how much would be taken out if he earned $220?

Equation: $\frac{12}{100} = \frac{x}{220}$

$100x = 2640$

$x = 26.4$

Answer: $26.40 would be taken out of $220.

A survey has shown that 20% of teenagers think they are overweight. In a class of 325, how many would this be?

Equation: $\frac{20}{100} = \frac{x}{325}$

$100x = 325 \cdot 20$

$x = 65$

Answer: 65 students

The price of gasoline was 96¢ a gallon before it went up 25%. How much did it go up? x = amount gas went up

Equation: $\frac{96+x}{96} = \frac{125}{100}$

$100(96+x) = 96 \cdot 125$

$9600 + 100x = 12{,}000$

$100x = 2400$

$x = 24$

Answer: It went up 24¢ a gallon.

Book 7, Page 31

Solve each problem using a proportion.

A jacket is on sale for $68, which is 80% of its original price. What was the price before the sale?

Equation: $\frac{68}{x} = \frac{80}{100}$

$80x = 68 \cdot 100$

$80x = 6800$

$x = 85$

Answer: It was $85.

Thea would like to leave a 15% tip. Her bill was $5.40. How much should she leave?

Equation: $\frac{15}{100} = \frac{x}{540}$

$100x = 15 \cdot 540$

$100x = 8100$

$x = 81$

Answer: She should leave 81¢.

The number of students at Plainview High went from 520 to 624 in just one year. This year's student body is what percent of last year's?

Equation: $\frac{624}{520} = \frac{x}{100}$

$520x = 624 \cdot 100$

$520x = 62400$

$x = 120$

Answer: 120% of last year's

Ted read on a cereal box that a serving contains 3 grams of protein, which is 4% of the protein a person should eat each day. According to this, how many grams of protein should a person eat daily?

Equation: $\frac{3}{x} = \frac{4}{100}$

$4x = 300$

$x = 75$

Answer: 75 grams per day

Read the problem above again. This time figure out what percent last year's student body was of this year's.

Equation: $\frac{520}{624} = \frac{x}{100}$

$624x = 520 \cdot 100$

$624x = 52000$

$x = 83\frac{1}{3}$

Answer: $83\frac{1}{3}$% of this year's

In the science lab Shad found that a candle used 60 out of 270 ml of trapped air before going out. What percent of the air did the candle use?

Equation: $\frac{60}{270} = \frac{x}{100}$

$\frac{270x}{270} = \frac{6000}{270}$

$x = 22.22$

Answer: 22.22% of the air

Key to Algebra – ANSWERS

Book 7, Page 32

Time Problems

How long *would* it take Jack and Jan working together? We can write an equation to find the answer. All we have to do is to think about how much of the work each one can do in an hour.

	Total Time	*Amount Done in an Hour*
Jack	6 hours	$\frac{1}{6}$
Jan	5 hours	$\frac{1}{5}$
together	x hours	$\frac{1}{x}$

Think: If we add the amounts each one does in an hour, we should get the amount they do together in an hour.

Equation: $\frac{1}{6} + \frac{1}{5} = \frac{1}{x}$

$$30x\left(\frac{1}{6} + \frac{1}{5}\right) = \left(\frac{1}{x}\right)30x$$

$$5x + 6x = 30$$

$$11x = 30$$

$$x = \frac{30}{11} = 2\frac{8}{11}$$

Answer: Jack was right! They can do the job together in less than three hours.

Book 7, Page 33

Write an equation for each problem. Then solve it to find the answer.

Jon delivers all his newspapers in 3 hours. It takes his trainee 4 hours to cover the route. How long will it take them together if they start from opposite ends?

Equation: In one hour:
Jon delivers $\frac{1}{3}$
Trainee delivers $\frac{1}{4}$
Together they deliver $\frac{1}{x}$

$$12x\left(\frac{1}{3} + \frac{1}{4}\right) = \frac{1}{x} \cdot 12x$$

$$\frac{12x}{3} + \frac{12x}{4} = \frac{12x}{x}$$

$$4x + 3x = 12$$

$$7x = 12$$

$$x = \frac{12}{7}$$

$$x = 1\frac{5}{7}$$

Answer: It will take them $1\frac{5}{7}$ hours.

Rex eats a bag of dog food in 12 days. Duke goes through a bag in 6 days. How long will one bag feed the two of them?

Equation: In one day:
Rex eats $\frac{1}{12}$ bag.
Duke eats $\frac{1}{6}$ bag.
Together they eat $\frac{1}{x}$ bag.

$$12x\left(\frac{1}{12} + \frac{1}{6}\right) = \left(\frac{1}{x}\right)12x$$

$$x + 2x = 12$$

$$3x = 12$$

$$x = 4$$

Answer: 4 days

Tracy can mow the lawn with a power mower in 20 minutes. Hai can mow it with a hand mower in 50 minutes. How long will it take them if they cooperate?

Equation: In one minute:
Tracy mows $\frac{1}{20}$
Hai mows $\frac{1}{50}$
Together they mow $\frac{1}{x}$

$$100x\left(\frac{1}{20} + \frac{1}{50}\right) = \left(\frac{1}{x}\right)100x$$

$$5x + 2x = 100$$

$$7x = 100$$

$$x = \frac{100}{7}$$

$$x = 14\frac{2}{7}$$

Answer: $14\frac{2}{7}$ minutes

The old duplicator could produce an issue of the newsletter in 2 hours. The new one takes 40 minutes. How long will it take if both machines are used?

Equation: In one minute:
Old dupl. makes $\frac{1}{120}$ issue
New one makes $\frac{1}{40}$ issue
Together they make $\frac{1}{x}$

$$120x\left(\frac{1}{120} + \frac{1}{40}\right) = \left(\frac{1}{x}\right)120x$$

$$x + 3x = 120$$

$$4x = 120$$

$$x = 30$$

Answer: 30 minutes

Book 7, Page 34

Problems About Rational Numbers

Make up an equation for each problem. Then solve the equation to get the answer.

"I'm thinking of a number. If you add half of this number and a third of this number, you will get 35. What is my number?"

Equation: $\frac{x}{2} + \frac{x}{3} = 35$

$$6\left(\frac{x}{2} + \frac{x}{3}\right) = 35 \cdot 6$$

$$3x + 2x = 210$$

$$5x = 210$$

$$x = \frac{210}{5} = 42$$

Answer: 42

"I'm thinking of an integer. When half of this integer is added to a third of the next integer, the result is 7. What is the integer?"

Equation: $\frac{x}{2} + \frac{x+1}{3} = 7$

$$6\left(\frac{x}{2} + \frac{x+1}{3}\right) = 7 \cdot 6$$

$$3x + 2(x+1) = 42$$

$$3x + 2x + 2 = 42$$

$$5x + 2 = 42$$

$$5x = 40$$

$$x = 8$$

Answer: 8

"I'm thinking of a number. If I add this number to its reciprocal, I get 2. What is the number?"

Equation: $x + \frac{1}{x} = 2$

$$x\left(x + \frac{1}{x}\right) = 2 \cdot x$$

$$x^2 + 1 = 2x$$

$$x^2 - 2x + 1 = 0$$

$$(x-1)(x-1) = 0$$

$$x - 1 = 0$$

$$x = 1$$

Answer: 1

"If I take a fourth of a number away from half the number, I get 10. What is the number?"

Equation: $\frac{x}{2} - \frac{x}{4} = 10$

$$4\left(\frac{x}{2} - \frac{x}{4}\right) = 10 \cdot 4$$

$$2x - x = 40$$

$$x = 40$$

Answer: 40

Diophantus was a famous mathematician of ancient Greece. A legend says there was a number puzzle on his tomb — a puzzle that you can solve the way you solved the other problems on this page.

"This tomb holds Diophantus. For one sixth of his life he was a boy, for one twelfth, a youth. After one seventh more he married and five years later had a son. The son lived only half as long as his father, and died four years before him." How long did Diophantus live?

x = Diaph. age
$\frac{x}{2}$ = son's age

Equation: $84\left(\frac{x}{6} + \frac{x}{12} + \frac{x}{7} + 5 + \frac{x}{2} + 4\right) = (x)\,84$

$$14x + 7x + 12x + 420 + 42x + 336 = 84x$$

$$75x + 756 = 84x$$

$$756 = 9x$$

$$x = 84$$

Answer: 84 years

Book 7, Page 35

Answers to Written Work

①

$$\frac{5}{8} + \frac{2}{5} = \frac{5 \cdot 5}{8 \cdot 5} + \frac{2 \cdot 8}{5 \cdot 8} = \frac{25}{40} + \frac{16}{40} = \frac{41}{40}$$

$$\frac{5}{8} - \frac{2}{5} = \frac{5 \cdot 5}{8 \cdot 5} - \frac{2 \cdot 8}{5 \cdot 8} = \frac{25}{40} - \frac{16}{40} = \frac{9}{40}$$

$$\frac{5}{8} \cdot \frac{2}{5} = \frac{1}{4}$$

$$\frac{5}{8} \div \frac{2}{5} = \frac{5}{8} \times \frac{5}{2} = \frac{25}{16}$$

$$\frac{2}{x} + \frac{4}{y} = \frac{2y}{xy} + \frac{4x}{yx} = \frac{2y+4x}{xy}$$

$$\frac{2}{x} - \frac{4}{y} = \frac{2y}{xy} - \frac{4x}{yx} = \frac{2y-4x}{xy}$$

$$\frac{2}{x} \cdot \frac{4}{y} = \frac{8}{xy}$$

$$\frac{2}{x} \div \frac{4}{y} = \frac{2}{x} \cdot \frac{y}{4} = \frac{y}{2x}$$

$$\frac{a}{a+3} + \frac{3}{a+3} = \frac{a+3}{a+3} = 1$$

$$\frac{a}{a+3} - \frac{3}{a+3} = \frac{a-3}{a+3}$$

$$\frac{a}{a+3} \cdot \frac{3}{a+3} = \frac{3a}{(a+3)^2} \text{ or } \frac{3a}{a^2+6a+9}$$

$$\frac{a}{a+3} \div \frac{3}{a+3} = \frac{a}{a+3} \cdot \frac{a+3}{3} = \frac{a}{3}$$

$$\frac{4}{n+2} + \frac{4}{n-2} = \frac{4(n-2)}{(n+2)(n-2)} + \frac{4(n+2)}{(n-2)(n+2)} = \frac{4n-8}{(n+2)(n-2)} + \frac{4n+8}{(n+2)(n-2)} = \frac{8n}{(n+2)(n-2)}$$

$$\frac{4}{n+2} - \frac{4}{n-2} = \frac{4(n-2)}{(n+2)(n-2)} - \frac{4(n+2)}{(n-2)(n+2)} = \frac{4n-8}{(n+2)(n-2)} - \frac{4n+8}{(n+2)(n-2)}$$

$$= \frac{4n-8}{(n+2)(n-2)} + \frac{-4n+{}^{-}8}{(n+2)(n-2)} = \frac{-16}{(n+2)(n-2)} \text{ or } \frac{-16}{n^2-4}$$

Key to Algebra – ANSWERS

Book 7, Page 35 (cont.)

Answers to Written Work

$$\frac{4}{n+2}\cdot\frac{4}{n-2}=\frac{16}{n^2-4}$$

$$\frac{4}{n+2}\div\frac{4}{n-2}=\frac{4}{n+2}\cdot\frac{n-2}{4}=\frac{n-2}{n+2}$$

$$\frac{a}{5b}+\frac{b}{10a^2}=\frac{a\cdot 2a^2}{5b\cdot 2a^2}+\frac{b\cdot b}{10a^2\cdot b}=\frac{2a^3}{10a^2b}+\frac{b^2}{10a^2b}=\frac{2a^3+b^2}{10a^2b}$$

$$\frac{a}{5b}-\frac{b}{10a^2}=\frac{a\cdot 2a^2}{5b\cdot 2a^2}-\frac{b\cdot b}{10a^2\cdot b}=\frac{2a^3}{10a^2b}-\frac{b^2}{10a^2b}=\frac{2a^3-b^2}{10a^2b}$$

$$\frac{a}{5b}\cdot\frac{b}{10a^2}=\frac{1}{50a}$$

$$\frac{a}{5b}\div\frac{b}{10a^2}=\frac{a}{5b}\cdot\frac{10a^2}{b}=\frac{2a^3}{b^2}$$

$$\frac{x+9}{x-2}+\frac{x-4}{x+3}=\frac{(x+9)(x+3)}{(x-2)(x+3)}+\frac{(x-4)(x-2)}{(x+3)(x-2)}=\frac{x^2+12x+27}{(x-2)(x+3)}+\frac{x^2-6x+8}{(x+3)(x-2)}=\frac{2x^2+6x+35}{(x-2)(x+3)}$$

$$\frac{x+9}{x-2}-\frac{x-4}{x+3}=\frac{(x+9)(x+3)}{(x-2)(x+3)}-\frac{(x-4)(x-2)}{(x+3)(x-2)}=\frac{x^2+12x+27}{(x-2)(x+3)}-\frac{x^2-6x+8}{(x+3)(x-2)}$$

$$=\frac{x^2+12x+27}{(x-2)(x+3)}+\frac{-x^2+6x-8}{(x+3)(x-2)}=\frac{18x+19}{(x-2)(x+3)}$$

$$\frac{x+9}{x-2}\cdot\frac{x-4}{x+3}=\frac{x^2+5x-36}{x^2+x-6}$$

$$\frac{x+9}{x-2}\div\frac{x-4}{x+3}=\frac{x+9}{x-2}\cdot\frac{x+3}{x-4}=\frac{x^2+12x+27}{x^2-6x+8}$$

② $$\frac{2}{3}+\frac{4}{5}-\frac{3}{4}=\frac{2\cdot 20}{3\cdot 20}+\frac{4\cdot 12}{5\cdot 12}-\frac{3\cdot 15}{4\cdot 15}=\frac{40}{60}+\frac{48}{60}-\frac{45}{60}=\frac{43}{60}$$

$$\frac{5}{4}+\frac{10}{x}+\frac{3}{2x^2}=\frac{5\cdot x^2}{4\cdot x^2}+\frac{10\cdot 4x}{x\cdot 4x}+\frac{3\cdot 2}{2x^2\cdot 2}=\frac{5x^2}{4x^2}+\frac{40x}{4x^2}+\frac{6}{4x^2}=\frac{5x^2+40x+6}{4x^2}$$

$$\frac{1}{cd}+\frac{2}{c^2d}+\frac{3}{cd^2}=\frac{1\cdot cd}{cd\cdot cd}+\frac{2\cdot d}{c^2d\cdot d}+\frac{3\cdot c}{cd^2\cdot c}=\frac{cd}{c^2d^2}+\frac{2d}{c^2d^2}+\frac{3c}{c^2d^2}=\frac{cd+2d+3c}{c^2d^2}$$

$$\frac{x}{y}+x+y=\frac{x}{y}+\frac{xy}{y}+\frac{y^2}{y}=\frac{x+xy+y^2}{y}$$

$$\frac{1}{6x}-\frac{1}{8x}-\frac{1}{10x}=\frac{1\cdot 20}{6x\cdot 20}-\frac{1\cdot 15}{8x\cdot 15}-\frac{1\cdot 12}{10x\cdot 12}=\frac{20}{120x}-\frac{15}{120x}-\frac{12}{120x}=\frac{-7}{120x}$$

Book 7, Page 35 (cont.)

Answers to Written Work

$$7-\frac{2}{x+1}=\frac{7(x+1)}{x+1}-\frac{2}{x+1}=\frac{7x+7}{x+1}+\frac{-2}{x+1}=\frac{7x+5}{x+1}$$

$$3x-\frac{1}{x}=\frac{3x\cdot x}{x}-\frac{1}{x}=\frac{3x^2}{x}+\frac{-1}{x}=\frac{3x^2-1}{x}$$

$$\frac{4}{x-5}-\frac{5}{x+4}=\frac{4(x+4)}{(x-5)(x+4)}-\frac{5(x-5)}{(x+4)(x-5)}=\frac{4x+16}{(x-5)(x+4)}-\frac{5x-25}{(x+4)(x-5)}$$

$$=\frac{4x+16}{(x-5)(x+4)}+\frac{-5x+25}{(x+4)(x-5)}=\frac{-x+41}{(x+4)(x-5)}$$

$$\frac{x}{x+2}+\frac{3}{x^2+2x}=\frac{x}{x+2}+\frac{3}{x(x+2)}=\frac{x\cdot x}{x(x+2)}+\frac{3}{x(x+2)}=\frac{x^2+3}{x(x+2)}$$

$$\frac{3}{x-1}-\frac{2x+4}{x^2-1}=\frac{3}{x-1}-\frac{2x+4}{(x+1)(x-1)}=\frac{3(x+1)}{(x-1)(x+1)}-\frac{2x+4}{(x+1)(x-1)}$$

$$=\frac{3x+3}{(x-1)(x+1)}+\frac{-2x-4}{(x+1)(x-1)}=\frac{\cancel{x-1}}{(x+1)\cancel{(x-1)}}=\frac{1}{x+1}$$

$$\frac{x+9}{x^2-25}+\frac{4}{x+5}=\frac{x+9}{(x+5)(x-5)}+\frac{4(x-5)}{(x+5)(x-5)}=\frac{x+9}{(x+5)(x-5)}+\frac{4x-20}{(x+5)(x-5)}$$

$$=\frac{5x-11}{(x+5)(x-5)}$$

$$\frac{x+1}{2x+10}-\frac{x-1}{3x+15}=\frac{x+1}{2(x+5)}-\frac{x-1}{3(x+5)}=\frac{3(x+1)}{3\cdot 2(x+5)}-\frac{2(x-1)}{2\cdot 3(x+5)}=\frac{3x+3}{6(x+5)}-\frac{2x-2}{6(x+5)}$$

$$=\frac{3x+3}{6(x+5)}+\frac{-2x+2}{6(x+5)}=\frac{\cancel{x+5}}{6\cancel{(x+5)}}=\frac{1}{6}$$

$$\frac{2}{x^2-1}+\frac{1}{(x-1)^2}-\frac{1}{(x+1)^2}=\frac{2}{(x+1)(x-1)}+\frac{1}{(x-1)^2}-\frac{1}{(x+1)^2}$$

$$=\frac{2(x-1)(x+1)}{(x+1)(x-1)(x-1)(x+1)}+\frac{1(x+1)^2}{(x-1)^2(x+1)^2}-\frac{1(x-1)^2}{(x+1)^2(x-1)^2}$$

$$=\frac{2(x^2-1)}{(x+1)^2(x-1)^2}+\frac{x^2+2x+1}{(x-1)^2(x+1)^2}-\frac{x^2-2x+1}{(x+1)^2(x-1)^2}$$

$$=\frac{2x^2-2}{(x+1)^2(x-1)^2}+\frac{x^2+2x+1}{(x-1)^2(x+1)^2}+\frac{-x^2+2x-1}{(x+1)^2(x-1)^2}$$

$$=\frac{2x^2+4x-2}{(x+1)^2(x-1)^2}=\frac{2(x^2+2x-1)}{(x+1)^2(x-1)^2}$$

Book 7, Page 35 (cont.)

Answers to Written Work

$$\frac{x}{x^2+2x-48}+\frac{2}{x^2-4x-12}=\frac{x}{(x+8)(x-6)}+\frac{2}{(x-6)(x+2)}$$

$$=\frac{x(x+2)}{(x+8)(x-6)(x+2)}+\frac{2(x+8)}{(x-6)(x+2)(x+8)}$$

$$=\frac{x^2+2x}{(x+8)(x-6)(x+2)}+\frac{2x+16}{(x-6)(x+2)(x+8)}$$

$$=\frac{x^2+4x+16}{(x+8)(x-6)(x+2)}$$

③ Terry was right. Sandy should have found a common denominator for the two fractions as follow:

$$\frac{5}{x}+\frac{1}{2x}=\frac{5\cdot 2}{x\cdot 2}+\frac{1}{2x}=\frac{10}{2x}+\frac{1}{2x}=\frac{11}{2x}$$

④ a) $\frac{117}{300}=\frac{x}{100}$

$300x=117\cdot 100$

$\frac{\cancel{300}x}{\cancel{300}}=\frac{11700}{300}$

$x=\frac{117}{3}=39$

39% of the students would like to be like their parents.

b) $\frac{183}{300}=\frac{x}{500}$

$300x=183\cdot 500$

$\frac{\cancel{300}x}{\cancel{300}}=\frac{91500}{300}$

$x=\frac{915}{3}=305$

305 out of 500 would not like to be like their parents

Book 7, Page 36

Practice Test

Add. Simplify your answer if you can.

$$\frac{7}{5}+\frac{8}{5}=\frac{\cancel{15}^{3}}{\cancel{5}_{1}}=3 \qquad \frac{y}{2x}+\frac{3}{2x}=\frac{y+3}{2x}$$

$$\frac{6}{x+5}+\frac{5}{x+5}=\frac{11}{x+5} \qquad \frac{x-2}{x+2}+\frac{x+6}{x+2}=\frac{2x+4}{x+2}=\frac{2\cancel{(x+2)}}{\cancel{x+2}}=2$$

$$\frac{8}{a^2}+\frac{1\cdot a}{a\cdot a}=\frac{8}{a^2}+\frac{a}{a^2}=\frac{8+a}{a^2} \qquad \frac{q\cdot 3}{q\cdot p}+\frac{4\cdot p}{q\cdot p}=\frac{3q+4p}{pq}$$

$$\frac{x}{1}+\frac{x}{4}=\frac{4x}{4}+\frac{x}{4}=\frac{4x+x}{4}=\frac{5x}{4} \qquad \frac{3}{1}+\frac{2}{x-5}=\frac{3(x-5)}{x-5}+\frac{2}{x-5}=\frac{3x-15}{x-5}+\frac{2}{x-5}=\frac{3x-13}{x-5}$$

$$\frac{5\cdot 2}{5(x+10)}+\frac{3}{\underset{5(x+10)}{\cancel{5x+50}}}=\frac{10}{5(x+10)}+\frac{3}{5(x+10)}=\frac{13}{5(x+10)}$$

$$\frac{12}{\underset{(x-3)(x+3)}{\cancel{x^2-9}}}+\frac{2(x-3)}{(x+3)(x-3)}=\frac{12}{(x-3)(x+3)}+\frac{2x-6}{(x+3)(x-3)}=\frac{2x+6}{(x+3)(x-3)}=\frac{2\cancel{(x+3)}}{\cancel{(x+3)}(x-3)}=\frac{2}{x-3}$$

$$\frac{(x-1)\,x}{(x-1)(x+4)}+\frac{2(x+4)}{(x-1)(x+4)}=\frac{x^2-x}{(x-1)(x+4)}+\frac{2x+8}{(x-1)(x+4)}=\frac{x^2+x+8}{(x-1)(x+4)}$$

Subtract. Simplify your answer if you can.

$$\frac{14}{9}+\frac{-2}{9}=\frac{12}{9}=\frac{4}{3} \qquad \frac{2}{a-1}+\frac{-5}{a-1}=\frac{-3}{a-1} \qquad \frac{x+4}{7}+\frac{-x+5}{7}=\frac{9}{7}$$

$$\frac{5}{2n}-\frac{2\cdot 2}{n\cdot 2}=\frac{5}{2n}+\frac{-4}{2n}=\frac{1}{2n} \qquad \frac{3x\cdot 3}{3x\cdot 2x}-\frac{4}{6x^2}=\frac{9x}{6x^2}-\frac{4}{6x^2}=\frac{9x-4}{6x^2}$$

$$\frac{3x}{2x+1}+\frac{-x}{2x+1}=\frac{2x}{2x+1}$$

$$\frac{(x+3)\,x}{(x+3)(x-3)}-\frac{6x}{\underset{(x+3)(x-3)}{\cancel{x^2-9}}}=\frac{x^2+3x}{(x+3)(x-3)}+\frac{-6x}{(x+3)(x-3)}=\frac{x^2-3x}{(x+3)(x-3)}=\frac{x\cancel{(x-3)}}{(x+3)\cancel{(x-3)}}=\frac{x}{x+3}$$

$$\frac{(x+1)(x+2)}{(x+1)(x-4)}-\frac{3(x-4)}{(x+1)(x-4)}=\frac{x^2+3x+2}{(x+1)(x-4)}+\frac{-3x+12}{(x+1)(x-4)}=\frac{x^2+14}{(x+1)(x-4)}$$

Key to Algebra – ANSWERS

Book 7, Page 37

Solve each equation.

$4\left(\frac{x}{2} - \frac{x}{4}\right) = 6 \cdot 4$
$2x - x = 24$
$x = 24$

$\frac{x}{5} = \frac{3}{7}$
$\frac{7x}{7} = \frac{15}{7}$
$x = \frac{15}{7}$

$2x \cdot \left(\frac{1}{x} + \frac{3}{2x}\right) = \frac{1}{2} \cdot 2x$
$2 + 3 = x$
$5 = x$
$x = 5$

$\frac{x-1}{x} = \frac{x}{4}$
$4(x-1) = x^2$
$4x - 4 = x^2$
$0 = x^2 - 4x + 4$
$0 = (x-2)(x-2)$
$x - 2 = 0$
$x = 2$

Make up an equation for each problem. Then solve the equation to get the answer.

If two inches on a map represents 25 miles, how many miles would 7 inches represent?

Equation: $\frac{2}{25} = \frac{7}{x}$
$\frac{2x}{2} = \frac{175}{2}$
$x = 87.5$

Answer: 87.5 miles

45 of the 75 seniors have jobs. What percent of the seniors are working?

Equation: $\frac{45}{75} = \frac{x}{100}$
$75x = 45 \cdot 100$
$75x = 4500$
$x = 60$

Answer: 60% are working

Pat can shelve a cart full of books in the library in 24 minutes and Kevin can do it in 30 minutes. How long will it take them if they work together?

Equation: $120x\left(\frac{1}{24} + \frac{1}{30}\right) = \frac{1}{x} \cdot 120x$
$2 \cdot 2 \cdot 2 \cdot 3 \cdot 5$ $2 \cdot 2 \cdot 2 \cdot 3$ $3 \cdot 5 \cdot 2$
$5x + 4x = 120$
$9x = 120$
$x = 13\frac{1}{3}$

Answer: $13\frac{1}{3}$ minutes

"I'm thinking of a number. If you add half of this number to a sixth of this number, you will get 12. What is the number?"

Equation: $6\left(\frac{x}{2} + \frac{x}{6}\right) = 12 \cdot 6$
$3x + x = 72$
$4x = 72$
$x = 18$

Answer: 18

Key to Algebra – NOTES

Key to Algebra – NOTES

Key to Algebra – NOTES

Key to Algebra® workbooks

Book 1: Operations on Integers
Book 2: Variables, Terms, and Expressions
Book 3: Equations
Book 4: Polynomials
Book 5: Rational Numbers
Book 6: Multiplying and Dividing Rational Expressions
Book 7: Adding and Subtracting Rational Expressions
Book 8: Graphs
Book 9: Systems of Equations
Book 10: Square Roots and Quadratic Equations

Answers and Notes for Books 1–4
Answers and Notes for Books 5–7
Answers and Notes for Books 8–10
Reproducible Tests for Books 1–10

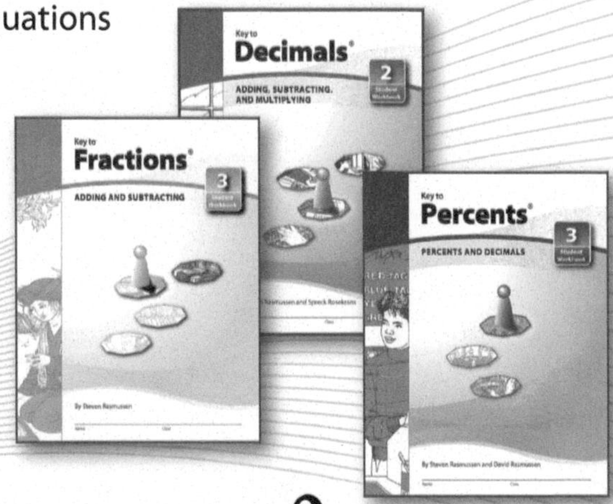

Also available in the Key to...® series

Key to Fractions®
Key to Decimals®
Key to Percents®
Key to Geometry®
Key to Measurement®
Key to Metric Measurement®

The KeyTo TRACKER™
The Online Companion to the *Key to...*® Workbook Series

Save time, improve learning, and monitor student progress with The Key To Tracker, the online companion to the Key to... workbooks for fractions, decimals, percents, and algebra.

Learn more: www.keypress.com/keyto

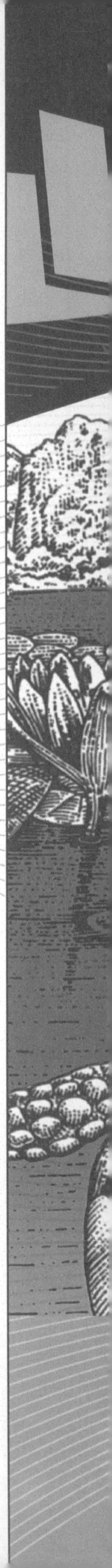

Key Curriculum Press
INNOVATORS IN MATHEMATICS EDUCATION

ISBN 978-1-55953-014-9
9 781559 530149 90000